Edge Computing Acceleration

IEEE Press
445 Hoes Lane
Piscataway, NJ 08854

Edge Computing Acceleration

From 5G to 6G and Beyond

Patrick Hung, Hongwei Kan, and Greg Knopf

IEEE ComSoc
IEEE Communications Society
The ComSoc Guides to Communications Technologies
Nim K. Cheung, Series Editor

IEEE Press
WILEY

Published by John Wiley & Sons, Inc., Hoboken, New Jersey.
Published simultaneously in Canada.

Library of Congress Cataloging-in-Publication Data Applied for:

Hardback ISBN: 9781119813842

Cover Design: Wiley
Cover Image: © DrPixel/Getty Images

Set in 9.5/12.5pt STIXTwoText by Straive, Chennai, India

Contents

About the Authors

Patrick Hung received his PhD in electrical engineering from Stanford University and served as a consulting assistant professor there from 2003 to 2009. He has been recognized as a CBI overseas scholar in the United Kingdom and a Taishan scholar in China. Patrick co-founded several high-tech companies and is the co-director of the CALAS Laboratory at City University of Hong Kong. He also serves as the vice chair of the IEEE Hong Kong Computer Society.

Hongwei Kan is the General Manager of the Emerging Technologies Research Laboratory at Tsinghua Unigroup. His research interests include xPU and chiplet design, heterogeneous computing platforms, emerging computer architectures, and artificial intelligence. Before joining Tsinghua Unigroup, Hongwei was the Chief Architect at Inspur Group, overseeing high-performance computing and storage platforms. Professor Kan is an IEEE Senior Member.

Greg Knopf is the Senior Director of Server Customer Engineering at AMD, where he oversees the development of AMD EPYC-based server platforms from concept to launch. His research interests include software/hardware co-development, cloud-native technologies, at-scale debug, and hardware acceleration technologies in artificial intelligence and other datacenter applications. Before joining AMD, Greg was the Senior Director of Engineering at Intel, leading multiple worldwide design teams.

Foreword (Professor Ray Cheung)

As a university professor in computer engineering, teaching system-on-chip, and a good friend to the authors, I am pleased to introduce you to *Edge Computing Acceleration: From 5G to 6G and Beyond*. Authored by Patrick, Hongwei, and Greg, this book is a comprehensive guide to the rapidly evolving world of edge computing and its intersection with 5G and 6G technologies.

We are at a pivotal point in the technological landscape where the convergence of edge computing and advanced communication networks is set to redefine our digital experiences. This book is a testament to the authors' deep understanding and expertise in high-performance computing, semiconductor chip design, digital security, and communication networks from academic and industry perspectives.

The book is structured into nine insightful chapters, each devoted to a critical aspect of edge computing acceleration. It begins with an introduction to the fundamentals of edge and custom computing, leading us through the complexities of 5G and 6G networks. The authors expertly discuss high-level synthesis (HLS) as a tool for accelerating hardware design and delve into the essential coding design principles for ensuring security in edge computing systems.

This book will serve as an invaluable resource and handbook in my senior-year undergraduate and postgraduate classes. It is designed to equip students, researchers, and practitioners with the knowledge and skills necessary to navigate and contribute to the field of edge computing acceleration. The blend of theoretical concepts and practical examples fosters a deep understanding of the subject matter and encourages innovation in this exciting domain.

I have had the distinct pleasure of co-teaching a university course on system-on-chip design with Patrick, one of the authors of this book. His deep understanding of the subject matter and his ability to convey complex concepts in an accessible manner have made him a favorite among students. Patrick is at the forefront of academic study, constantly pushing boundaries and exploring new areas in the field. His extensive experience in academia and his direct engagement with students have given him a unique perspective on the needs and challenges faced by learners in this rapidly evolving field.

As we embark on this journey of exploration and learning together, I am confident that *Edge Computing Acceleration: From 5G to 6G and Beyond* will guide you through the complexities and opportunities of this transformative era in computing and communication. I wholeheartedly recommend this book to anyone interested in the future of edge computing and the evolution of 5G and 6G technologies.

Associate Provost, City University of Hong Kong
Chair, IEEE Hong Kong Section

Professor Ray Cheung

Foreword (Raghu Nambiar)

The industry landscape is evolving at an unprecedented pace, driven by rapid technological progress and a shift toward smarter and more efficient edge devices. The advent of 5G and the anticipation of 6G are fundamentally reshaping the digital infrastructure landscape. *Edge Computing Acceleration: From 5G to 6G and Beyond*, authored by industry experts Patrick Hung, Hongwei Kan, and Greg Knopf, offers a profound exploration of the technologies shaping our interconnected tomorrow.

The subjects explored in this book hold personal significance for me. In my current role as the lead for software and solution engineering at AMD, and in my previous position as the CTO of the server business at Cisco overseeing emerging technology solutions, I have engaged closely with these topics. This book aligns seamlessly with my market insights and aspirations. I am convinced it offers valuable insights for both knowledge seekers and practitioners.

In my career journey, I have personally witnessed the transformative power of collaboration in driving technological progress. This book is a shining example of such collaboration, as its authors bring together a wealth of knowledge from both academic and industry perspectives. Over the years, I have had the privilege of closely collaborating with Greg Knopf, an industry leader who has played a crucial role in the successful deployment of millions of servers in data centers across the globe. I wholeheartedly extend my congratulations and appreciation to him and the co-authors for their outstanding teamwork, seamlessly blending academic insights with practical industry perspectives.

As you delve into this enlightening guide, I urge you to envision the boundless potential of these technologies. The capability to process data at the edge, in data centers, and throughout the network unlocks unparalleled opportunities across diverse industries. Truly, we are living in an exhilarating era in computing and telecommunications. My wish is that this book not only serves as a comprehensive roadmap but also sparks inspiration to innovate and push the limits of what is achievable.

Corporate Vice President, AMD
San Francisco Bay Area

Raghu Nambiar

Preface

We are standing at the cusp of a technological revolution, where the confluence of 5G, 6G, and edge computing is reshaping the landscape of digital communication and computation. It is a pivotal moment that demands a comprehensive understanding of these technologies and their potential to accelerate the performance and capabilities of edge computing. In this context, we have endeavored to present *Edge Computing Acceleration: From 5G to 6G and Beyond*, a book that aims to bridge the gap between theoretical concepts and practical applications in this rapidly evolving domain.

The genesis of this book lies in our diverse expertise across high-performance computing, semiconductor chip design, digital security, and communication networks. Patrick Hung, an expert in computer architecture and SOC design; Hongwei Kan, an architect of heterogeneous acceleration platforms; and Greg Knopf, a specialist in CPU and HPC platforms, have collaboratively crafted a comprehensive resource that addresses the multifaceted aspects of edge computing acceleration in the era of 5G and 6G.

This book is structured into nine chapters, each focusing on a critical aspect of edge computing acceleration. The journey begins with an introduction to the fundamentals of edge and custom computing and its significance in the context of 5G and 6G networks. We then delve into the technical intricacies of 5G and 6G, exploring their architecture, capabilities, and the role they play in enhancing edge computing. High-level synthesis (HLS) is discussed as a pivotal tool for accelerating hardware design, followed by an examination of coding design principles for ensuring security and integrity in edge computing systems.

The hardware and software architectures form the backbone of efficient edge computing systems, and we dedicate two chapters to dissect these components and their interplay. As the potential of edge computing is realized through its applications, we explore the "Killer Applications" that are set to transform industries. The transition from concept to production is a critical phase, and we provide insights and guidelines for navigating this process successfully. Lastly, the "The Road Ahead" chapter offers a forward-looking perspective on the future developments in edge computing and the evolving landscape of 5G and 6G technologies.

Intended for use in a semester-long graduate-level course, this book is designed to equip students, researchers, and practitioners with the knowledge and skills necessary to navigate and contribute to the field of edge computing acceleration. Through a blend of theoretical concepts and practical examples, we aim to foster a deep understanding of the subject matter and inspire innovation in this exciting domain.

As we embark on this journey together, we hope that *Edge Computing Acceleration: From 5G to 6G and Beyond* serves as a valuable resource, guiding you through the complexities and opportunities of this transformative era in computing and communication.

March 2024 *Patrick Hung, Hongwei Kan, and Greg Knopf*

Acknowledgment (Patrick Hung)

I am profoundly grateful to everyone who has contributed to the creation of this book. This work represents a collaborative endeavor to delve into the cutting-edge of edge acceleration technology and its evolution through the eras of 5G, 6G, and beyond.

In this book, we tackle the challenges of the 5G era, where an array of artificial intelligence (AI) and robotic applications necessitate hardware acceleration through accelerators such as graphic processing units (GPUs), field programmable gate arrays (FPGAs), and data processing units (DPUs). Regrettably, the proprietary nature of most commercial solutions makes it challenging to grasp the bigger picture. This project was conceived to systematically educate students and engineers in edge acceleration technology, filling the knowledge gap and fostering a deeper understanding of this crucial field.

I extend my heartfelt appreciation to my co-authors, Greg Knopf and Hongwei Kan, for their invaluable contributions, insights, and collaborative spirit. Greg's expertise in CPUs and servers, coupled with Hongwei's proficiency in high-performance computing, has been instrumental in this project. Working alongside such knowledgeable and dedicated individuals has been both an honor and a privilege.

Special acknowledgment goes to Professor Qin Huang from Bei Hang University, who provided most of the materials for Chapter 4. Professor Huang is dedicated to advanced coding research and has the remarkable ability to simplify complex mathematical concepts, significantly enhancing the quality and accessibility of this book.

I am deeply indebted to Kam Sum and Abdurrashid Ibrahim Sanka, two dear friends who dedicated countless hours to reviewing every chapter. Without their assistance, this book would not have been possible. We are also grateful to our editor and anonymous reviewers for their constructive criticism and guidance, which have greatly improved the quality and clarity of our work.

We owe a debt of gratitude to our colleagues and peers in the academic and industry communities for their invaluable insights, feedback, and contributions, which have enriched the content and depth of this book. I would like to express my profound gratitude to Professor Michael Flynn, whose support and guidance were instrumental in my research on computer architecture. Special thanks also go to our research teams and assistants, whose dedication, hard work, and attention to detail have been instrumental in bringing this project to fruition.

I would like to express my heartfelt thanks to my wife, Jane, and my daughter, Madeline, for their constant support and encouragement. Their belief in me has been a constant source of inspiration.

Our appreciation extends to the various organizations and institutions that provided the necessary resources, infrastructure, and financial support for our research. Their support has been essential in enabling us to explore and push the boundaries of edge computing technology.

Lastly, we acknowledge the readers and the broader community interested in edge computing and its future developments. We hope that this book serves as a valuable resource and inspires further innovation and exploration in the field.

Thank you all for being a part of this journey.

Sincerely,

March 2024 *Patrick Hung*

Acknowledgment (Greg Knopf)

I would like to express my immense gratitude to everyone who contributed to the creation of this book and to the many influential people in my life who provided support and inspiration throughout my academic and professional journey.

First, I would like to thank my co-authors, Dr. Patrick Hung and Hongwei Kan. I am deeply grateful to Patrick for initiating the project, graciously inviting me to participate, and steadfastly leading it through to completion as we juggled our various responsibilities in life and work. Patrick has been a great friend and mentor since we first collaborated on server platform development and memory technology in the mid-2000s. We have had nearly two decades of deep and exciting conversations about industry trends and emerging technologies, often stretching our dinners in the San Francisco Bay Area or Hong Kong until restaurants are practically pushing us out the door. I am equally grateful to our co-author Hongwei Kan, who brought extremely valuable insights and contributions to the project from his extensive background in high-performance computing.

I am tremendously grateful to my wife, Adrienne, for her constant support and encouragement along the way. She helped me find time and space to focus and pursue this endeavor with my colleagues. I am also immensely grateful for our lovely daughter Heidi, my never ending source of inspiration.

I would also like to thank my lifelong friends Allen White, John Dering, David Dieruf, William Morrow III, David Wolford, Chad Hartley, and Pat Evans for always motivating and encouraging each other in our various pursuits.

At AMD, I have had the honor to work with renowned engineering leaders who are actively shaping the computing industry today, and I feel very fortunate to have the opportunity to learn from them. Dr. Lisa Su, AMD CEO, has been a spectacular leader and constant source of inspiration. It has also been an immense pleasure and learning experience to work closely with Mark Papermaster, AMD CTO, as we have grown the EPYC CPU product line and business.

I would also like to thank Jay Kirkland for his many years of leadership, collaboration, and friendship. Jay's support and partnership over many years helped to create numerous opportunities and relationships that played an important role in the path to developing this book.

Finally, I want to acknowledge and thank all of the readers of this book and the broader community. We are living in an extremely exciting time in the world of computing, and we hope that this book helps to fuel new innovations and developments in the industry.

Thank you,
D. Greg Knopf

Boulder, Colorado
May 2024

Part I

Introduction

1

Introduction

The fifth-generation (5G) and the upcoming sixth-generation (6G) wireless networks are poised to be the most significant and transformative technology in the coming decade, disrupting numerous industries ranging from energy, agriculture, manufacturing to transportation, retail, healthcare, entertainment, and financial services. Surprisingly and unbeknownst to most outsiders, one of the key technology enablers is the new computer architecture and software design model, in addition to the wireless communications technology [1, 2].

The main objective of this book is to help university students and professional engineers understand the 5G/6G edge computing architecture and to describe step by step how to unleash the full 5G/6G potential using custom edge computing technologies.

This book is divided into five parts:

- Part 1 (Introduction):
  - Chapters 1 and 2 introduce the concept of edge and custom computing and provide an overview of 5G/6G technologies.
- Part 2 (Theory):
  - Chapters 3 and 4 discuss how to use high-level synthesis (HLS) and coding theory to realize secure edge acceleration with high performance.
- Part 3 (Architecture):
  - Chapters 5 and 6 elaborate the 5G/6G hardware and software acceleration architecture.
- Part 4 (Applications):
  - Chapters 7 and 8 describe a few 5G/6G killer applications with acceleration technology including some practical development strategies.
- Part 5 (Future Roadmap):
  - Chapter 9 discusses the road ahead in the coming decade.

Edge Computing Acceleration: From 5G to 6G and Beyond, First Edition.
Patrick Hung, Hongwei Kan, and Greg Knopf.

1.1 Introducing 5G and Internet of Everything

With 20 times higher bandwidth and 10 times reduction in network latency, fifth-generation (5G) technologies enable many new applications, such as remote surgery and autonomous driving. These incredible applications require both high bandwidth and extremely low latency. To realize these useful applications, we need high-performance communication and computing platforms demanding different kinds of acceleration technologies, known as heterogeneous acceleration architecture (Figure 1.1).

Although the fundamental wireless technology – wireless spectral efficiency – has only increased three times from 4G/LTE to 5G, the area traffic capacity is 100 times larger. This is made possible mainly by building many more 5G cell towers, each covering a smaller area and supporting a wider spectrum. A typical 4G/LTE cell tower may have a range of 10 km, but a typical 5G tower may have a range of 500 m or less.

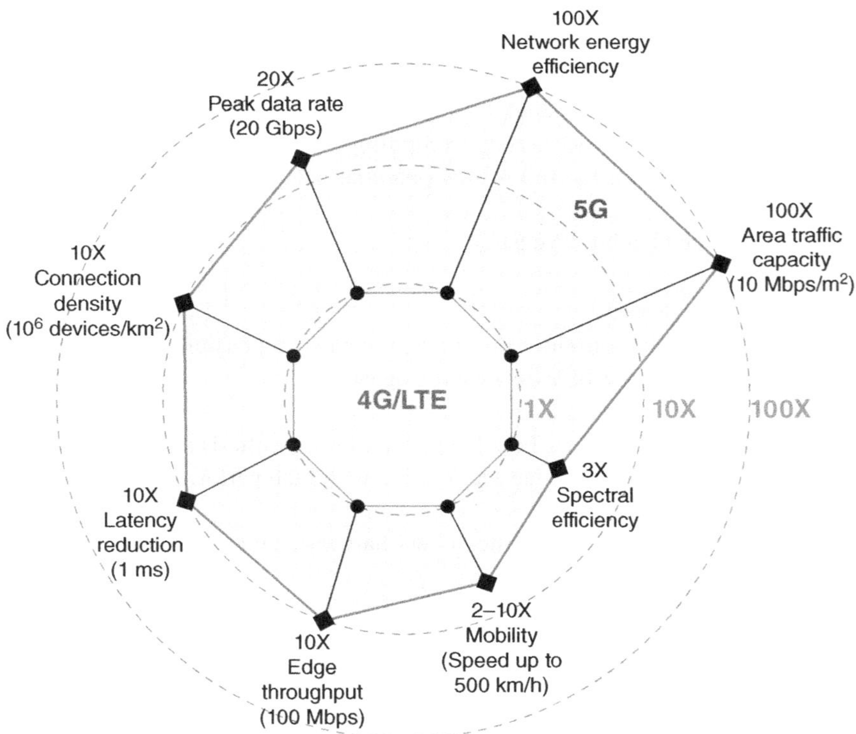

Figure 1.1 Comparison between 4G and 5G features and performance metrics.

The 3rd-Generation Partnership Project (3GPP) is responsible for driving the 5G telecommunications standards, which are detailed in the 3GPP Specifications Release 15 (NR Phase 1) and Release 16 (NR Phase 2) documents. The 3GPP 5G network architecture defines the network entities based on their functions and nature (control and data planes) [3–6].

There are three technical specification groups (TSGs) within the 3GPP:

- **TSG SA** (services and systems aspects) focuses on the overall architectures and services.
- **TSG CT** (core network and terminals) focuses on the core network (CN) architecture and terminal interfaces.
- **TSG RAN** (radio access networks) focuses on the radio transmission and its technical requirements.

The TSG SA defines three different sets of requirements for new 5G usages:

- **Enhanced Mobile Broadband (eMBB):** a new requirement that defines higher data rates, traffic and connection densities, and user mobility.
- **Massive Machine-Type Communications (mMTC):** a new requirement that supports very high traffic densities of devices.
- **Ultra-Reliable Low Latency Communications (URLLC):** a new requirement that provides very low latency and very high communications service availability (Figure 1.2).

Figure 1.2 compares the bandwidth and latency requirements of some 4G and 5G applications. The typical 4G network latency is 10–20 ms, while the minimum 5G network latency can go as low as 1 ms. The maximum 4G network bandwidth is around 100 Mbps, but the 5G network bandwidth can go up to 1 Gbps. Some new mobile applications, such as mobile telepresence, may only require higher network bandwidth. On the other hand, most disruptive applications, such as autonomous driving and smart energy grid, may demand very short network latencies.

Although human beings may not perceive any difference between 5 and 50 ms, this difference is so important in many industrial robotic systems. Indeed, an interconnection of machine systems – called the Internet of Everything (IoE) – can take full advantage of the three new 5G features (eMBB, mMTC, and URLLC) [7, 8] (Figure 1.3).

In 2012, Cisco Systems extended the concept of the Internet of Things (IoT) and coined the term "the Internet of Everything (IoE)," representing a networked connection of people, processes, data, and things. It goes beyond simple machine-to-machine (M2M) communications, forming a network of networks connecting all data, technologies, processes, and people [9–12].

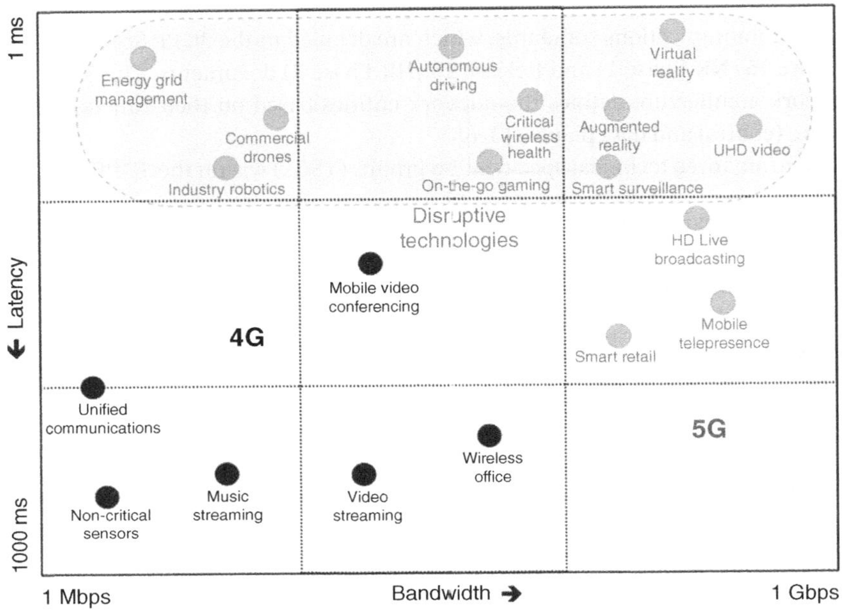

Figure 1.2 Latency and bandwidth requirements of some typical 4G and 5G applications.

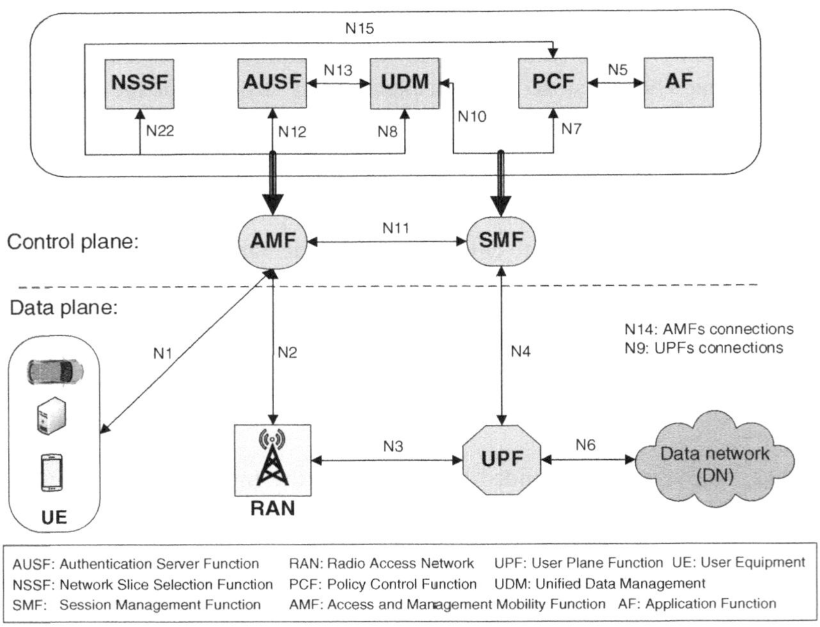

Figure 1.3 3GPP 5G network architecture.

Figure 1.4 Ubiquitous Internet of Everything (IoE) devices.

Today many IoE devices are adopting a tethered communication scheme – using Wi-Fi or Bluetooth to first connect to a smartphone or an access point – to communicate with other IoE devices. Unfortunately, tethered communication is slow, unreliable, and insecure, limiting the performance of these applications [13, 14].

The 5G/6G wireless network enables IoE devices to adopt an untethered communication scheme, thus allowing each device to have a secure channel connecting to a radio access network (RAN) with guaranteed network latency, bandwidth, and security (Figure 1.4).

According to Juniper Research, IoE devices reached 46 billion units by 2021. Figure 1.4 depicts a framework of different connected devices and things. The top IoE applications include smart factories, smart buildings, smart grids, connected gaming and entertainment, smart vehicles, remote healthcare, remote education, and connected marketing and advertisement.

These ubiquitous IoE devices have some essential characteristics:

- **Resource Limitations:** Many IoE devices are limited by size, weight, and power (SWaP) and do not have sufficient resources to process or store information locally. Accordingly, the information is offloaded to a remote server for storage and processing.
- **Network Latency:** Extremely low network latency is needed for time-critical applications. Industrial control systems can only tolerate delays of the order of milliseconds. Autonomous vehicles and virtual reality applications also have

similar latency requirements. The network latency can be excessive if the servers are in a remote cloud.

- **Network Bandwidth:** A large IoE network can generate a huge amount of real-time data. For example, 120,000 CCTV devices transmitting 1080p video to a remote cloud may generate 1 Tbps traffic. If the servers are in a remote cloud, the 1 Tbps network data can cause serious traffic congestion in the Internet backbone.
- **Security and Reliability:** Many critical applications, such as energy grids and smart factories, must be ultra-secure and reliable. Due to the multi-hop nature of the Internet backbone, it is probably impossible to guarantee ultra-security and reliability with a remote cloud server.
- **Reconfigurability:** After deployment, an IoE system oftentimes requires future security patches and performance updates. While it is easier to update a generic software architecture, it is difficult to update a high-performance hardware architecture. This is an important reason why custom computing is desirable for 5G computing architecture.

1.2 Edge Computing Architecture

Although 5G can provide large bandwidths and low latencies, if the other end of the communication is far away, these parameters cannot be guaranteed. In particular, the communication latency is limited by the speed of light. For example, a distance of 4000 km – between Los Angeles and New York – has at least 20 ms delay on fiber optics [15–17].

In reality, the actual latencies, including all electronic and electro-optic delays, are much longer. Table 1.1 shows the typical round-trip latencies from Tokyo to some target cities.

Table 1.1 Round-trip time (RTT) from Tokyo to overseas cities.

Target city	Round-trip time from Tokyo (ms)
Cape Town	360
New York	176
London	218
Los Angeles	128
Paris	236
San Paulo	275
Shanghai	67
Sydney	114

Source: Adapted from http://Wondernetwork.com.

It is important to note that a saving of 20 ms – migrating from a 4G network to a 5G network – is not meaningful if the application involves a network round-trip time of 360 ms. We must keep this in mind when we implement a remote application, such as remote surgery or Internet gaming. The remote server cannot be too far away, or the underlying algorithm must be able to hide the latency from the users.

As much as possible, the IoE servers should be placed close to the wireless networks to optimize the network security, reliability, bandwidth, and latency. This requirement gives rise to the emerging **edge computing** architectures, where servers are located at the edge of the Internet near the wireless devices. Therefore, edge computing is a distributed computing paradigm or a model where the processing and storage elements are placed close to the sensing device (data source) for improved latency, throughput, security, and several other benefits. The traditional method sends the sensor data to the cloud for processing and storage. As shown in Figure 1.5, some edge computing nodes do not require the cloud because they are independent. The edge computing server is placed together (in the same housing) with the sensing device for many independent edge computing systems (Figure 1.5).

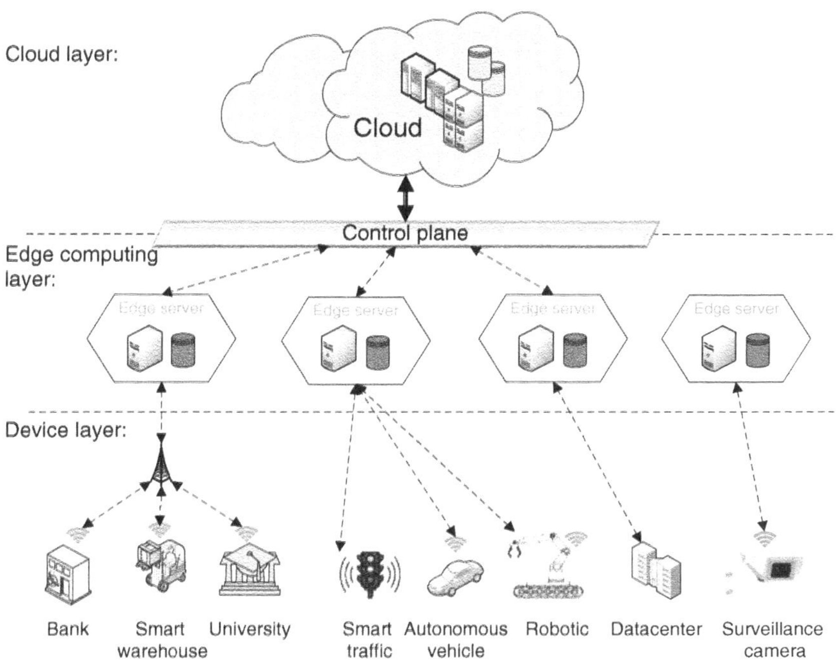

Figure 1.5 Edge versus cloud computing architecture.

1.2.1 Edge Versus Cloud Computing

Cloud computing, which has been a dominant enabler in enterprise IT delivery in the past decades, offers many key advantages [18]:

1) An enterprise can avoid the huge capital expenditure (CAPEX) of building a datacenter.
2) Cloud computing can lower operating expenses (OPEX) by exploiting economies of scale.
3) Cloud computing can minimize electricity costs by strategically locating servers in regions with the lowest electricity and air conditioning costs.

While cloud computing is a good value proposition for many enterprise applications, it is not a viable option for many time-critical IoE applications. Edge computing is a new paradigm in which substantial computing and storage resources are placed in close proximity to mobile devices or sensors [19–21] (Figure 1.6).

There have been many research activities and industry developments related to edge computing in the past decade:

- In early 2013, Nokia and IBM jointly introduced an edge computing platform called the Radio Applications Cloud Server (RACS).
- In 2014, a mobile edge computing standardization effort began under the auspices of the European Telecommunications Standards Institute (ETSI).

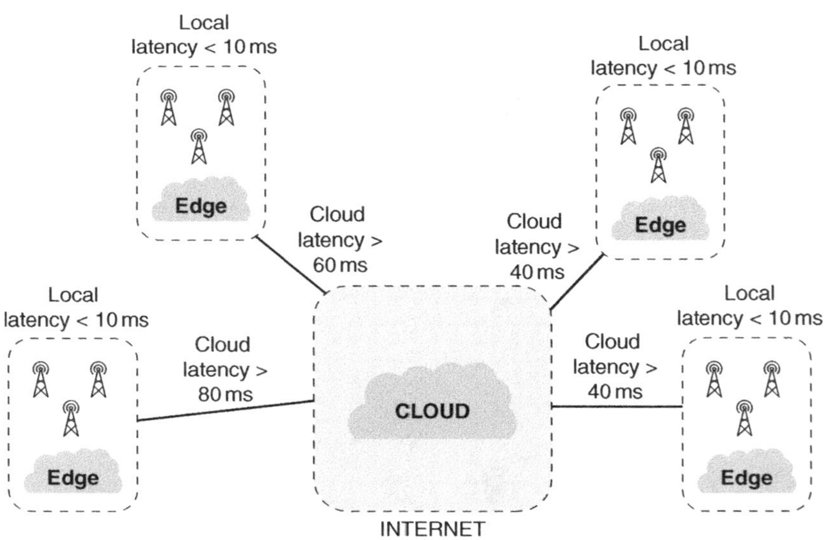

Figure 1.6 Edge versus cloud network latencies.

- In 2015, multiple mobile technology organizations – led by Vodafone, Intel, and Huawei – launched the Open Edge Computing Initiative (OEC).
- In late 2015, Cisco, Microsoft, Intel, Dell, and ARM, in partnership with Princeton University, created the OpenFog Consortium.

The origin of edge computing dates back to the 1990s, when Akamai introduced content delivery networks (CDNs) to accelerate web performance. A CDN uses nodes at the edge close to users to pre-fetch and cache web content. These edge nodes can also perform some content customizations, such as adding location-aware advertisements. The emerging edge computing architectures generalize and extend the CDN concept by leveraging cloud computing infrastructure within a 5G network.

1.2.2 Edge Design Options

Broadly speaking, there are two different architectural options for edge computing:

- **Mobile Edge Computing (MEC):** MEC consists of commercial off-the-shelf (COTS) computers located near the 5G RAN to reduce latency and improve context awareness. It is a standalone architecture that can function without a remote cloud.
- **Fog Computing (FC):** FC is a decentralized computing infrastructure with fog computing nodes (FCNs) placed between the end devices and a cloud. The FCNs are based on heterogeneous network elements including routers, switches, and gateways. Unlike MEC, FC is tightly linked to a cloud and cannot function without a remote cloud.

Both MEC and FC are computing and storage networks, which are built on top of 5G communications networks. Indeed, we are migrating from a simple 4G/LTE communications network to an integrated computing and communications network (Figure 1.7).

Interestingly, the 5G communications network has two similar layers:

- **Control Plane** (CP), also known as user plane function (UPF), consists of generic computers running a network operating system and controlling the network domain. The CP controls and manages the user equipment's access

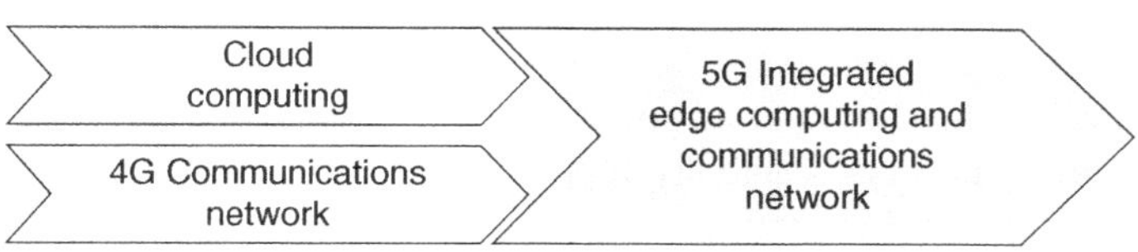

Figure 1.7 Edge computing is driving many 5G applications.

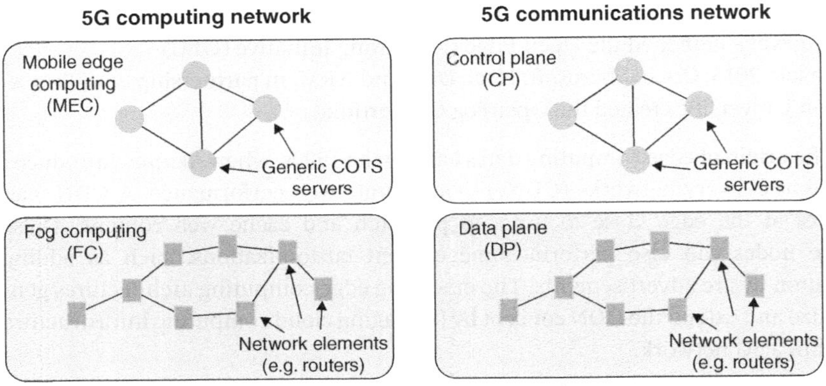

Figure 1.8 5G computing and communications networks.

to the network. It performs the access and mobility function (AMF), session management function (SMF), authentication server functions (AUSF), unified data management (UDM), and policy control functions.

- **Data Plane** (DP) consists of the configurable network devices within the network domain. The data plane interfaces with the radio access network (RAN) and the data network (DN), and then performs the UPF on the received user data. Therefore, the data plane performs tasks such as packet inspection and processing, routing, and ensuring the quality of service (QoS) (Figure 1.8).

In an integrated 5G communications and computing network, MEC and CP can likely share the same servers, while FC and DP can share the same network elements. Nevertheless, each implementation is different, and the mobile network operator (MNO) would determine the network architecture. Table 1.2 compares the two edge computing design options.

1.2.3 Key Benefits of Edge Computing

Edge computing provides the following key benefits:

- **Responsive Services:** Some mobile applications are very sensitive to the long latency in cloud computing. Applications such as autonomous driving, virtual reality (VR), augmented reality (AR), robotics, and energy grids can work better in the edge computing environment.
- **Network Scalability:** Edge computing can reduce massive ingress traffic to the cloud, thus improving network scalability. It processes massive raw data locally and can cut up to 99.9% of cloud traffic.

Table 1.2 Comparison of 5G private network deployment architectures.

Deployment architecture	RAN	DP	CP	MP	Perf. grade	Flexibility grade	Security grade	Investment efforts	Management simplicity	Resource efficiency
DA-A	On-prem	On-prem	On-prem	On-prem	Best	Best	Best	Worst	Worst	Worst
DA-B	On-prem	On-prem	On-prem	Remote	Best	Best	Medium	Worst	Medium	Medium
DA-C	On-prem	On-prem	Remote	Remote	Medium	Medium	Medium	Medium	Best	Best
DA-D	On-prem	Remote	Remote	Remote	Worst	Worst	Worst	Best	Best	Best

The table compares four different 5G private network deployment architectures (DA-A, DA-B, DA-C, and DA-D). DA-A is the most expensive and most flexible architecture, and DA-D is the cheapest and the least flexible architecture. DA-B and DA-C are between these two extremes.

- **Masking Cloud Outages:** Edge servers are independent and can function even without a remote cloud. When cloud outages are short, mobile devices may not even be aware of the outages.
- **Distributed Security:** Instead of storing all data in a centralized server, critical data are distributed in multiple edge computing servers. This creates a robust distributed security architecture.

1.3 Custom Computing

The 5G technologies have presented many market opportunities and challenges at the same time. Here are the seven most critical challenges (Figure 1.9).

We have discussed about the edge computing architecture in the last several sections. Although the standard edge computing architecture, using software-defined networking (SDN) and network function virtualization (NFV) [22, 23], can be cost-effective, scalable, flexible, and reliable, it may not be able to meet the stringent latency and bandwidth requirements. Thus, custom edge computing architecture can be used to dramatically speed up the 5G performance while maintaining the reconfigurability [24, 25].

1.3.1 Introduction to Custom Computing

Custom computing systems are special computer systems designed for specific applications, such as signal processing or database operations, and can be reconfigurable. These systems are desirable when the general-purpose system has undesirable features such as high hardware resource consumption, high latency, low throughput, and high-power consumption. Custom computing systems need to meet some SWaP requirements, usually employ special acceleration hardware elements, including:

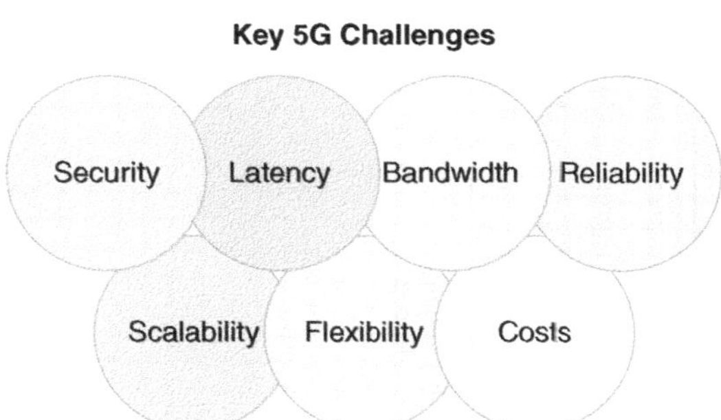

Figure 1.9 Key challenges to fully realize 5G potentials.

- **Field-Programmable Gate Arrays** (FPGA) chips, such as Xilinx Virtex-7.
- **Graphic Processing Unit** (GPU) chips, such as NVIDIA Titan V GPU.
- **System-on-Chip** (SOC), such as Google Tensor Processing Unit (TPU).

Custom computing is gaining popularity among all major cloud providers, thanks to the emergence of machine learning and big data applications. For example,

- Amazon EC2 F1 Instances and Alibaba ECS F3 Instances use FPGA-based PCI Express cards running on generic servers to enable the delivery of custom FPGA hardware acceleration.
- Amazon EC2 P2/P3/G3/G4 Instances and Alibaba EGS Instances use GPU-based PCI Express cards running on generic servers to enable the delivery of custom GPU acceleration.

By running the applications in hardware instead of software, custom computing can be used to improve their computational performance and reduce power consumption and costs. Unlike software algorithms, hardware implementations in FPGA and GPU are less susceptible to cyberattacks. As a result, custom computing architectures are ideal for implementing security elements, such as network firewalls and intrusion detection systems (IDSs).

1.3.2 5G/6G Security Concerns

Among the seven major challenges in the 5G era, security considerations are arguably the most critical concerns, receiving the most media attention in the last few years. Specifically, the 5G stakeholders are worried about the potential cyberattacks from foreign agents and malicious hackers [26–33].

Thanks to many emerging applications, the 5G networks are susceptible to many new threats. The common attack types include:

- **Denial of Service (DoS):** Attackers can temporarily disable the network services.
- **Network Message Modification:** Attackers can temporarily take over the network.
- **Data Corruption:** Attackers can erase important network data.
- **Eavesdropping:** Attackers can intercept confidential data from a 5G network.
- **Fraud:** Attackers can use network services at the expense of another subscriber.

In general, these cyberattacks may be originated from:

- Fake base stations.
- Rogue IoT devices, rogue base stations, and rogue network elements.
- Roaming networks.
- The Internet (Figure 1.10).

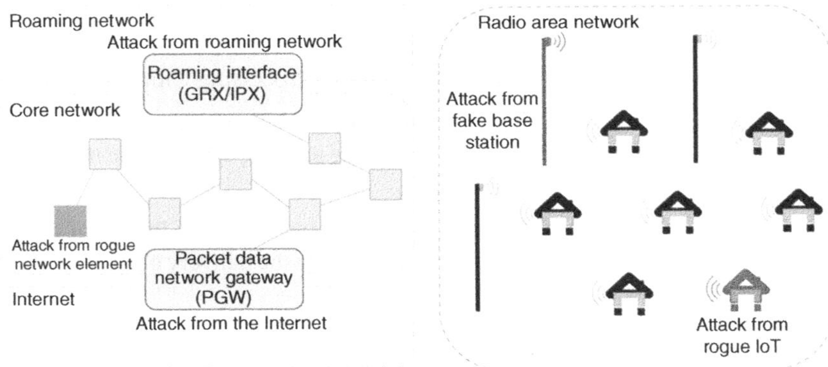

Figure 1.10 Many different ways to attack 5G network.

A DoS attack on a 5G/6G network is not only an inconvenient event but can also wreak havoc in an autonomous driving vehicle or a smart manufacturing factory. Accordingly, it is important to build a reliable 5G/6G network that can:

- Prevent cyberattacks.
- Detect cyberattacks quickly.
- Limit detrimental consequences of cyberattacks.

To prevent cyberattacks, it is useful to use custom computing technology to implement 5G high-performance security firewalls:

- **GTP firewalls** defend the network against GPRS tunneling protocol (GTP) based attacks. These cyberattacks may be initiated from RAN or roaming networks.
- **SGi/Gi firewalls** defend the network against TCP/IP attacks from the Internet. These attacks may exploit vulnerabilities in SMTP, NFS, FTP, Telnet, NTP, and HTTP.
- **Diameter firewalls** defend the network against diameter protocol-based attacks. Diameter protocol contains subscriber information, similar to the traditional SS7 protocol.

To detect cyberattacks, most enterprise networks today may already have IDSs. For example, Splunk IDS uses log data and history to identify network anomalies. However, Splunk IDS typically does not process the log data in real time. Accordingly, it is useful to use custom computing technology to implement 5G real-time IDSs.

To limit the detrimental consequences of a successful cyberattack, it is important not to use a centralized authentication and encryption server. Instead, it is useful to implement a distributed authentication and encryption architectures. These

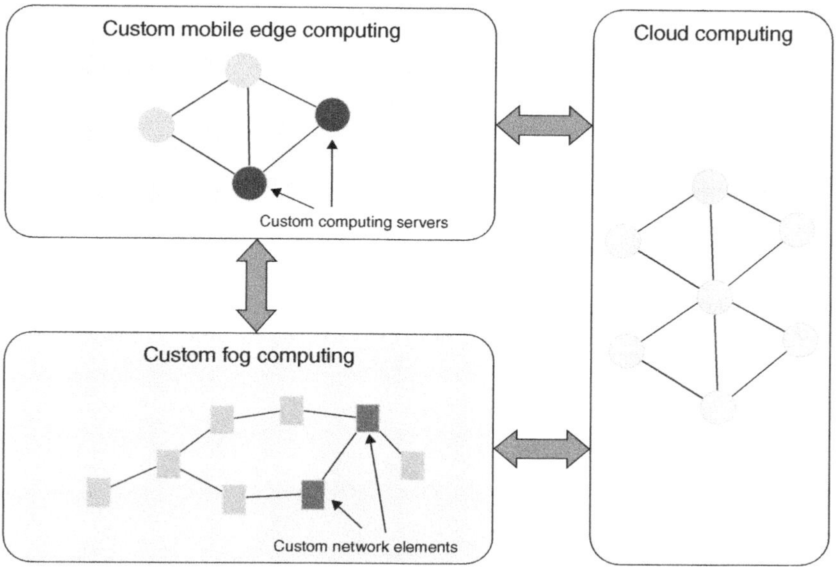

Figure 1.11 Employment of custom computing in mobile edge computing and fog computing. In both cases, custom computing can reduce network latency and power dissipation and enhance network bandwidth and digital security.

architectures can take advantage of custom computing technologies to implement complicated security algorithms.

1.3.3 Custom Edge Computing Cards

Custom edge computing systems can be employed in both mobile edge computing (control plane) and fog computing (user/data plane) (Figure 1.11).

In custom mobile edge computing architectures, one or more acceleration cards are added to COTS servers. These acceleration cards run the original software tasks in hardware, improving the performance by two or three orders of magnitude. At the same time, the power consumption can also be dramatically reduced (Figure 1.12).

In custom fog computing networks, special SOCs and FPGAs are added to the proprietary network computing elements. Oftentimes, the fog computing networks consist of proprietary embedded systems running non-standard operating systems. An emerging architecture is called the "Smart Network Interface" (or SmartNIC). Unlike a standard network interface card, a SmartNIC card may be FPGA based or SOC based and can be reconfigurable. A key feature of SmartNIC is minimizing the network latency (Figure 1.13).

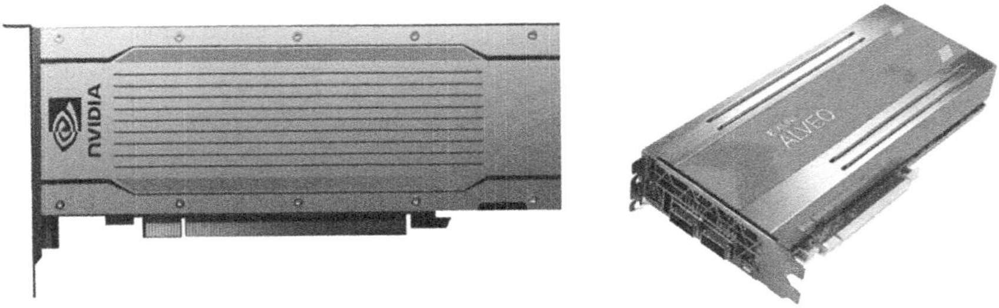

Figure 1.12 An NVIDIA Tesla T4 PCIe card and a Xilinx Alveo U200 PCIe card. In a 5G MEC or CP network, these acceleration cards can be used to increase performance and reduce overall power consumption.

Figure 1.13 Xilinx Alveo U25 SmartNIC acceleration card encompasses network, storage, and compute functions, providing a comprehensive SmartNIC platform.

1.4 Deployment Considerations

According to MTN Consulting, the revenues of MNOs have remained stagnant in the last few years with a 1.6% decline in 2019 even before the onset of the COVID-19 pandemic. On the other hand, building a 5G base station was three times more expensive than building a 4G base station, costing somewhere between US$30K and US$70K in 2019. To make the situation even worse, a typical 5G base station consumes twice or more the power of a 4G base station [34–36].

Construction and operation costs are the most critical barriers to the deployment of 5G network infrastructures. The conundrum is how we can unleash the full 5G potential without breaking the bank or taking excessive risks.

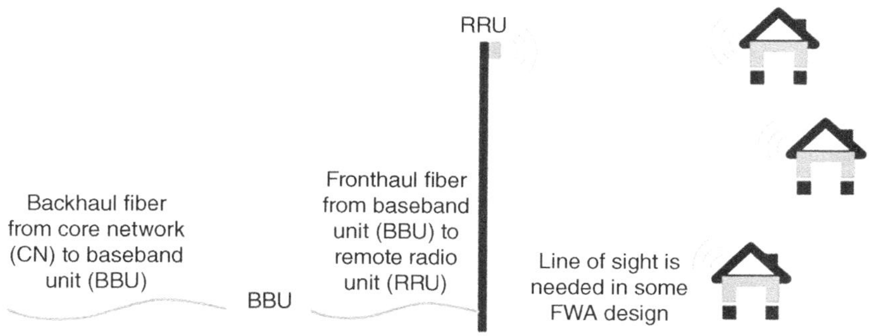

Figure 1.14 5G fixed wireless access (FWA) application.

What is the best way to deploy 5G? It depends on:

- Existing infrastructure.
- Use cases.
- Business models.

For example, some 5G equipment providers, including Ericsson and Huawei, actively promote the 5G fixed wireless access (FWA) market. Here, the household uses a 5G wireless router to provide Wi-Fi and LAN used within the house (Figure 1.14).

In this use case, there are two interesting points:

- First, the quality and reliability of the FWA network will depend on both the 5G wireless network and the fiber optic networks. Much of the 5G capital expenses are used to build the fiber optic infrastructure.
- Second, FWA is a good value proposition for many rural and sub-urban areas. However, FWA may not work well in some urban areas where fiber-to-the-premise (FTTP) is already commonplace. In urban areas, it makes sense to focus on other mobile applications.

In this section, we examine various 5G network deployment options and how they can affect the 5G network architecture.

1.4.1 5G/6G Cell Architecture

While the cell radius of a 4G network cell tower – called macrocell – varies between 2 and 40 km, a small cell may have a radius of less than 500 m [37–39]. A smaller 5G network cell can support higher traffic densities by taking advantage of:

- Higher frequency spectrums, which have shorter transmission ranges.
- More effective spectrum reuse across different cells.

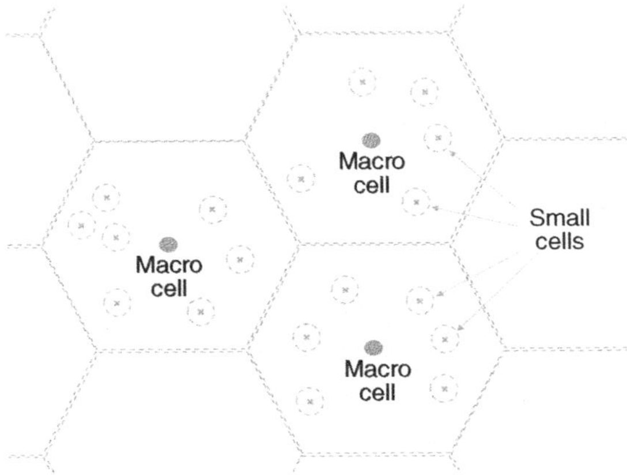

Figure 1.15 Small cells versus macrocells.

According to McKinsey's research, sites with traffic density above 0.5 petabytes per square kilometer per year will require a cell radius of less than 200 m. The traffic densities of many major cities are projected to reach 1–2 petabytes per square kilometer per year.

To leverage the existing infrastructure, most MNOs adopt a two-tier approach by [40, 41]:

- Upgrading the macrocells to support massive multi-input multi-output (mMIMO) protocols, with multiple antenna and beamforming technologies.
- Building small cells, as needed, within each macrocell (Figure 1.15).

These two approaches are complementary and somewhat independent. Several research papers discuss optimizing network resources. In reality, an MNO may decide a certain deployment option based on business considerations only.

There are different types of small cells:

- **Microcells:** with a radius of less than 2 km – are usually used to provide good wireless coverage within an in-building area, such as a train station or a shopping mall.
- **Picocells:** with a radius of less than 200 m – are usually used to cover a small area for a temporary event, such as a sporting event.
- **Femtocells:** with a radius of less than 10 m – are usually installed at home or in office.

Femtocells are often installed and maintained by end users, whereas microcells and picocells may be installed and maintained by either end users or MNOs. In all three cases, a 5G NR small cell must comply with 3GPP TS 38 series NG-RAN gNB specifications.

Here are some characteristics of a small cell:

- A small cell is connected to its 5G core network (5GC) by an NG interface, and the small cells are interconnected to each other by an Xn interface.
- Unlike a macrocell, a small cell may not support mMIMO with more than 16TX and 16RX.
- A small cell may support only 5G NR standards or both 4G/LTE and 5G NR standards.

1.4.2 5G/6G Private Network

In the previous section, we have discussed macrocells and small cells. Specifically, end users may install, maintain, and operate small cells. In this section, we will examine this specific market segment, called 5G private network.

There are a growing number of private 5G networks – aka non-public 5G networks – as we migrate from 4G/LTE to 5G. A private 5G network uses the same technology as public 5G networks but it allows an enterprise to manage its own 5G network to serve a limited geographic area with optimized services using dedicated equipment [42, 43].

A key component of a private 5G network is to provide:

- **Lower Latency:** In smart manufacturing, a private 5G network could be tailored for application delivery prioritization and to ensure predictable latency in all network traffic.
- **Higher Security:** A private 5G network could provide cellular-grade security, which can be used to keep sensitive data encrypted, authenticated, protected, and local to the premises.
- **Wider Coverage:** Many remote areas are not sufficiently covered by public cellular infrastructure. A private 5G network could be installed to provide wide coverage in agricultural fields, container ports, warehouses, or deep inside buildings.

Private 5G network use cases include:

- A smart factory can take advantage of a private 5G network to support services, such as predictive diagnostics, workforce management, machine automation, and factory safety management.
- A complex warehouse that fulfills pick-and-pack customer orders using robots and control software. A private 5G network is useful to help meet strict performance requirements needed to control countless fast-moving robots from a single base station.
- A large mine can use a private 5G network to support a complex heterogeneous system, comprising in-pit CCTV monitoring, intelligent earth sensing,

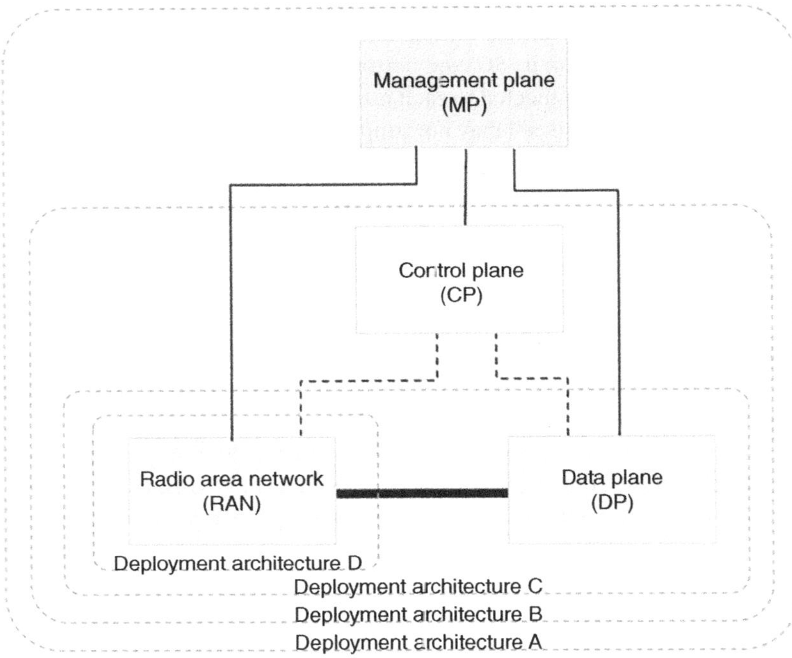

Figure 1.16 Deployment architectures A, B, C, and D.

telemetry, and many other monitoring services. This can help improve the reliability and safety of the mine.

There are many design alternatives in a private 5G network. Generally speaking, there are two main design aspects: deployment architecture and operation model [44–46]. Each design aspect has a few design options.

5G network architecture encompasses four main blocks:

- **Management Plane** (MP), aka management and orchestration layer, consists of network software managing a network domain.
- **Control Plane** (CP) consists of generic computers running a network operating system and controlling the network domain.
- **Data Plane** (DP), also known as the User Plane Function (UPF), consists of the configurable network devices within the network domain.
- Radio Access Network (RAN) consists of legacy and 5G NR (New Radio) interfaces supporting mMIMO and beamforming technologies (Figure 1.16).

In the most flexible private 5G network architecture (Deployment Architecture A), the above four blocks are all on-premise. While this deployment architecture is very secure and provides the best performance, it is also the most expensive. On the

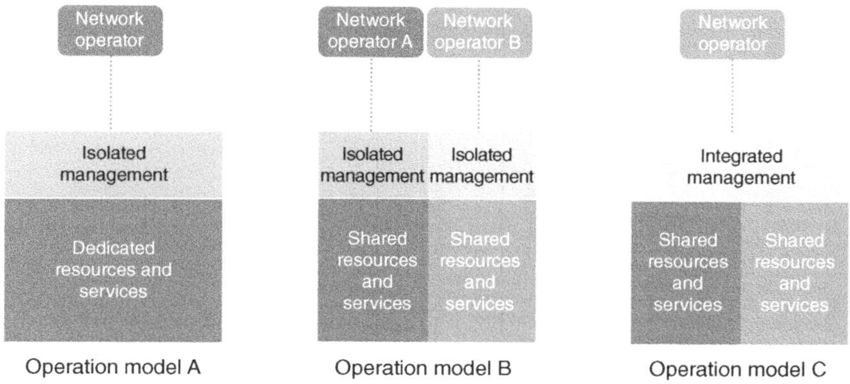

Figure 1.17 Different 5G operation models.

other extreme (Deployment Architecture D), only the RAN block is on-premise. The security level is the worst but the investment is the minimum (Table 1.2).

A private 5G network may adopt a certain operation model, independent from the deployment architecture discussed previously.

- First, the network infrastructure resources may be owned by a single user and used exclusively for a particular vertical application or are shared by multiple users and used for multiple vertical applications.
- Second, the network may be operated and managed independently from all other applications or by a third-party MNO that manages the other traffic at the same time (Figure 1.17 and Table 1.3).

1.4.3 Infrastructure Sharing

In the previous section, we have examined how a single private enterprise can launch its 5G wireless deployment using different private 5G/6G network architectures [42]. In this section, we will look at the 5G/6G network from a more global perspective.

The University of Cambridge and the Technical University of Madrid recently conducted research investigating various 5G deployment strategies in Britain. Specifically, the researchers focused on three questions:

- Should the MNOs increase or reduce the capital investment?
- Should the MNOs maintain the present infrastructure sharing?
- What is the reasonable 5G coverage expectation?

There are four MNOs sharing two macrocell networks in Britain:

- O_2/Telefonica and Vodafone are sharing one macrocell network.
- EE and Hutchison Three are sharing another macrocell network.

Table 1.3 Comparison of three 5G operation models.

Operation model	Operation and management	Network ownership	Usage simplicity	Service continuity	Liability protection	Security grade	Lock-in protection	Deployment flexibility
OM-A	Isolated	Dedicated	Worst	Worst	Worst	Best	Best	Best
OM-B	Isolated	Shared	Medium	Medium	Medium	Medium	Medium	Medium
OM-C	Integrated	Shared	Best	Best	Best	Worst	Worst	Worst

The table compares three different 5G private network operation models (OM-A, OM-B, and OM-C). OM-A is the most secure operating model; however, it requires the user to pay for its infrastructure and also to manage its network. This task may be too daunting for most users. OM-C is the cheapest and simplest operation model at the expense of less flexibility and security. OM-B is a compromise between OM-A and OM-C.

Table 1.4 Some CAPEX estimation for 5G deployment.

CAPEX model	CAPEX per year	CAPEX over 10 years
−25% model	£1.50 B	£15.0 B
−12.5% model	£1.75 B	£17.5 B
Baseline model	£2.00 B	£20.0 B
+12.5% model	£2.25 B	£22.5 B
+25% model	£2.50 B	£25.0 B

Together, the four MNOs are spending between £1.5 billion and £2.5 billion on capital expenses each year. For 5G deployment, the capital investment will primarily be spent on macrocell upgrades, small cell deployments, fiber backhaul, and core network upgrades. Using an average £2.0 billion spending per year as the baseline, the researchers projected the 5G coverage over the following years (Table 1.4 and Figure 1.18).

Based on the assumptions in this research, a large 5G investment is most effective in the first two years. Subsequently, the differences in between £1.5 billion per year and £2.5 billion per year are getting smaller and smaller, suggesting that an over-investment beyond the first two years has poor returns on investment (ROI).

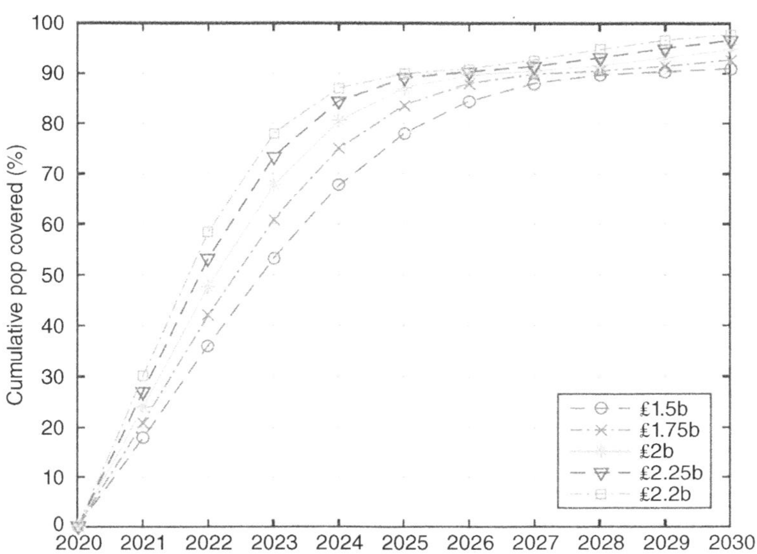

Figure 1.18 CAPEX requirements for various coverage models.

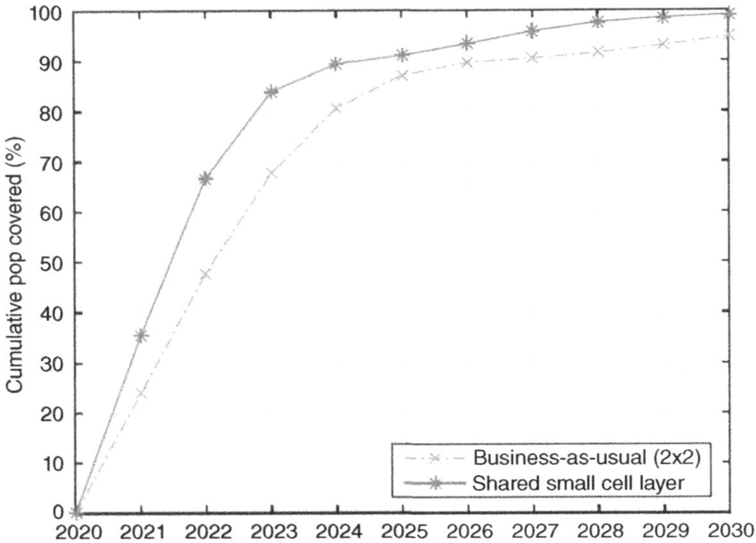

Figure 1.19 Optimizing coverage using shared small cell layer.

Second, the researchers investigated how the four MNOs can share the small cell infrastructure. This is made possible using the latest SDN/NFV technologies. The existing approach – with two independent small cell infrastructures – is too costly and ineffective. A single small cell infrastructure can quickly ramp up coverage in the first few years. More importantly, a single small cell infrastructure can reduce the deployment time by three or more years, when the coverage exceeds 90% (Figure 1.19).

Finally, the research suggests that providing 30 or 50 Mbps coverage for everyone in urban, suburban, and rural areas is too expensive. While it takes only £6 billion for the coverage of first 90% of the population, the remaining 10% will cost an additional £12 billion. A more sensible goal is to provide 50 Mbps for 90% of the population and 10 Mbps for the remaining 10% located in rural areas.

References

1 Wang, J., Jin, A., Shi, D. et al. (2017). Spectral efficiency improvement with 5G technologies: results from field tests. *IEEE Journal on Selected Areas in Communications* 35 (8): 1867–1875.

2 Holma, H., Toskala, A., and Nakamura, T. (2020). *5G Technology 3GPP New Radio*. Wiley.

3 Release description; Release 15. *3GPP Specification 21.915*, Alain Sultan, 2019.

4 Release description; Release 16. *3GPP Specification 21.916*, Alain Sultan, work-in-progress.

5 View on 5G architecture. Version 3.0. *5GPPP Architecture Working Group*, 4 February 2020. DOI 10.5281/zenodo.3265031.

6 5G spectrum access at 26 GHz and update on bands above 30 GHz. *Ofcom (UK)*, July 2017.

7 A. Sengupta, "Cellular terrestrial broadcast – physical layer evolution from 3GPP release 9 to release 16." *IEEE Transactions on Broadcasting*, Vol. 66, Issue 2, June 2020, pp. 459–470.

8 Pätzold, M. (2019). 5G is coming around the corner. *IEEE Vehicular Technology Magazine* 14 (1): 4–10.

9 Cisco Annual Internet Report (2018–2023) White Paper. *Cisco*, 9 March 2020.

10 Cisco Visual Networking Index: Global Mobile Data Traffic Forecast Update, 2017–2022. *Cisco*, February 2019.

11 Bradley, J., Loucks, J., Macaulay, J., and Noronha, A. (2013). Internet of Everything (IoE) Value Index. *Cisco*.

12 Target networks in 5G era, embracing mobile network 2020s. *Huawei*, 2017.

13 Albrecht, K. and Michael, K. (2013). Connected: to everyone and everything. *IEEE Technology and Society Magazine*, Winter.

14 Raj, A. (2018). Internet of everything: a survey based on architecture, issues and challenges. *2018 5th IEEE Uttar Pradesh Section International Conference on Electrical, Electronics and Computer Engineering (UPCON)*, 2–4 November 2018, Gorakhpur India, pp. 1–6. IEEE.

15 Satyanarayanan, M. (2017). The emergency of edge computing. *IEEE Computer* 50 (1): 30–39.

16 Dolui, K. and Datta, S.K. (2017). Comparison of fog computing, cloudlet and mobile edge computing. *IEEE GIoTS* 2017, 6–9 June 2017, CICG Geneva Switzerland.

17 Bright, J. (2019). Driving new business opportunities with edge computing and 5G. *Ovum*.

18 Edge computing made simple. Release 10.0. *Small Cell Forum*, December 2017.

19 Baccarelli, E., Naranjo, P.G.V., Scarpiniti, M. et al. (2017). Fog of everything: energy-efficient networked computing architectures, research challenges, and a case study. *IEEE Access* (5): 9882–9910.

20 Aazam, M. and Huh, E.N. (2016). Fog computing: the cloud-IoT/IoE middleware paradigm. *IEEE Potentials* 35 (3): 40–44.

21 Abdelwahab, S. (2014). Enabling smart cloud services through remote sensing: an internet of everything enabler. *IEEE Internet of Things Journal* 1 (3): 276–288.

22 Kreutz, D., Ramos, F.M.V., Verissimo, P.E. et al. (2015). Software-defined networking: a comprehensive survey. *Proceedings of the IEEE* 103 (1): 14–76.

23 Mijumbi, R., Serrat, J., Gorricho, J.L. et al. (2016). Network function virtualization: state-of-the-art and research challenges. *IEEE Communications Surveys & Tutorials* 18 (1): 236–262.

24 Todman, T.J., Constantinides, G.A., Wilton, S.J. et al. (2005). Reconfigurable computing: architectures and design methods. *IEE Proceedings-Computers and Digital Techniques* 152 (2): 193–207.

25 Sutton, A. (2018). 5G network architecture, design and optimisation. *IET 5G Conference*, 24 January 2018, London UK. https://www.academia.edu/35765795/5G_Network_Architecture_Design_and_Optimisation. Accessed online from Academia on 1 September 2024.

26 The Evolution of Security in 5G. 5G Americas' Board of Governors, July 2019.

27 Surridge, M., Correndo, G., Meacham, K., et al. (2018). Trust modeling in 5G mobile networks. *SecSoN '18: Proceedings of the 2018 Workshop on Security in Softwarized Networks: Prospects and Challenges*, 24 August 2018, Budapest Hungary.

28 Arfaoui, G., Bisson, P., Blom, R. et al. (2018). A security architecture for 5G networks. *IEEE Access* 6: 22466–22479.

29 Benzaïd, C. and Taleb, T. (2020). ZSM security: threat surface and best practice. *IEEE Network* 34 (3): 124–133.

30 M. Liyanage, Ahmed, I., Ylianttila, M., et al. (2015). Security for future software defined mobile network. *Paper presented at the 9th International Conference on NGMAST*, 9–11 September 2015, University of Cambridge UK.

31 Liyanage, M., Ahmad, I., Abro, A.B. et al. (2018). *A Comprehensive Guide to 5G Security*. Wiley.

32 Wheeler, T. (2019). If 5G is so important, why isn't it secure? *The New York Times* January 2019, USA.

33 Adam Janofsky, "The good news about 5G security." *The Wall Street Journal*, 10 2019, USA.

34 Lavallée, B. (2016). 5G wireless needs fiber, and lots of it. *Ciena*, November 2016.

35 Grijpink, F., Ménard, A., Sigurdsson, H., and Vucevic, N. (2018). *The Road to 5G: The Inevitable Growth of Infrastructure Cost*. McKinsey & Company.

36 Duncan Stewart and Paul Lee. (2019). Consumers, operators gear up for 5G. *Deloitte*, April 2019.

37 Deployment issues for enterprise small cells. Release 10.0. *Small Cell Forum*, December 2017.

38 Small cell siting challenges and recommendations. Release 10.0. *Small Cell Forum*, August 2018.

39 Walker, M. (2020). Operators facing power cost crunch. *MTN Consulting*.

40 5G small cell architecture and product definitions. *Small Cell Forum*, July 2020.

41 URLLC and network slicing in 5G enterprise small cell networks. Release 10.0. *Small Cell Forum*, February 2018.

42 Private LTE/5G networks: a primer for developers. *Qualcomm*, 2019.

43 Rostami, A. Private 5G networks for vertical industries: deployment and operation models. *2019 IEEE 2nd 5G World Forum*, 30 September–2 October 2019, Dresden Germany.

44 The Office of Communications Annual Report and Accounts (2019/2020). *Ofcom (UK)*, July 2020.

45 Oughton, E. and Frias, Z. (2018). The cost, coverage and rollout implications of 5G infrastructure in Britain. *Telecommunications Policy* 42: 636–652.

46 Romero-Gázquez, J., Moreno-Muro, F.J., Garrich, M. et al. (2019). A use case of shared 5G backhaul segment planning in an urban area. *Paper presented at ICTON* 2019, 9–13 July 2019, Angers France.

2

Overview of 5G and 6G

This chapter presents an overview of 5G development, covering its timeline, spectrum, and key characteristics. Specifically, we introduce the major technologies behind 5G and its key applications. In addition, we also briefly discuss 6G development and its potential applications in the future.

2.1 5G Timeline

The development of 5G started well before its first deployment in 2019. International Telecommunication Union (ITU), the United Nations' organization, manages wireless mobile telecommunications technology with the aim to connect the people of all over the world [1]. In 2015, ITU issued a document titled IMT Vision – framework and overall objectives of the future development of IMT for 2020 and beyond. IMT, which stands for International Mobile Telecommunications, encompasses the technical requirements and guidelines for global mobile communications. The document is a recommendation defining the framework and objectives and outlining three use cases of IMT 2020 in:

(i) Enhanced Mobile Broadband (eMBB),
(ii) Massive Machine Type Communications (mMTC), and
(iii) Ultra Reliable and Low Latency Communications (URLLC) [2].

The 3rd-Generation Partnership Project (3GPP), an industry group comprising key mobile telecommunication development organizations, produces technical specifications for the telecommunications industry [3] meeting the ITU's requirements. Previously, 3GPP published mobile communications standards for 2G, 3G, LTE and 4G technologies. LTE, which stands for "Long Term Evolution," was

Edge Computing Acceleration: From 5G to 6G and Beyond, First Edition.
Patrick Hung, Hongwei Kan, and Greg Knopf.

designed to significantly enhance the capacity and speed of networks beyond 3G. In 2015, 3GPP began the process of standardizing 5G technology.

3GPP released its first complete 5G standard (Release 15) in 2018, covering key topics of 5G New Radio (NR), 5G System (Phase 1), massive MTC and Internet of Things, vehicle-to-everything service, further LTE improvements, mobile communication system, and so on. This set of initial specifications covers the integration of non-standalone 5G systems with existing LTE networks. Subsequently, Release 16 in 2019 describes 5G and LTE improvements in areas such as Multiple-Input and Multiple-Output (MIMO) update, NR for unlicensed spectrum, power savings, and vertical expansion. Release 17 describes improvements over Release 16 in areas such as new use cases, new verticals, and ubiquitous connectivity. 3GPP is currently working on Release 18, focusing on 5G Advanced, which covers new technology components and adding intelligence such as machine learning into mobile networks. The timeline of major documents by ITU and 3GPP is shown in Figure 2.1.

2.2 5G Spectrum

The 5G standard defines its operating spectrum into two ranges: Frequency Range 1 (FR1) between 450 and 7125 MHz and Frequency Range 2 (FR2) between 24.25 and 52.6 GHz (mmWave), in 3GPP TS38.101 User Equipment (UE) radio transmission and reception [4, 5].

3GPP TS18.101 also defines the operating band in each frequency range. Details of the frequency bands are presented in Chapter 6. The choice of operating frequency band depends on the application scenario. In general, a low-frequency band (below 1 GHz) provides wider coverage (i.e., longer distance transmission) and can penetrate building walls when compared to a high-frequency band (e.g., over 20 GHz, also known as mmWave transmission). On the other hand, high-frequency radio signals offer high bandwidth, however, they more susceptible to attenuation over long distances and cannot penetrate building walls. Thus, high-frequency signals are good for high bandwidth but short-distance applications.

5G NR also supports carrier aggregation, which combines multiple frequency bands, resulting in a wider bandwidth. Through carrier aggregation, multiple simultaneous connections can be made. Such connections are useful for high-bandwidth applications such as augmented reality (AR) and virtual reality (VR). In addition, the use of supplementary carriers can enhance 5G network performance by adding additional frequency bands to the main carrier. When needed to avoid traffic congestion, a 5G network can use a supplementary carrier instead of the current uplink or downlink carrier, thus improving the network performance and efficiently using the available frequency bands.

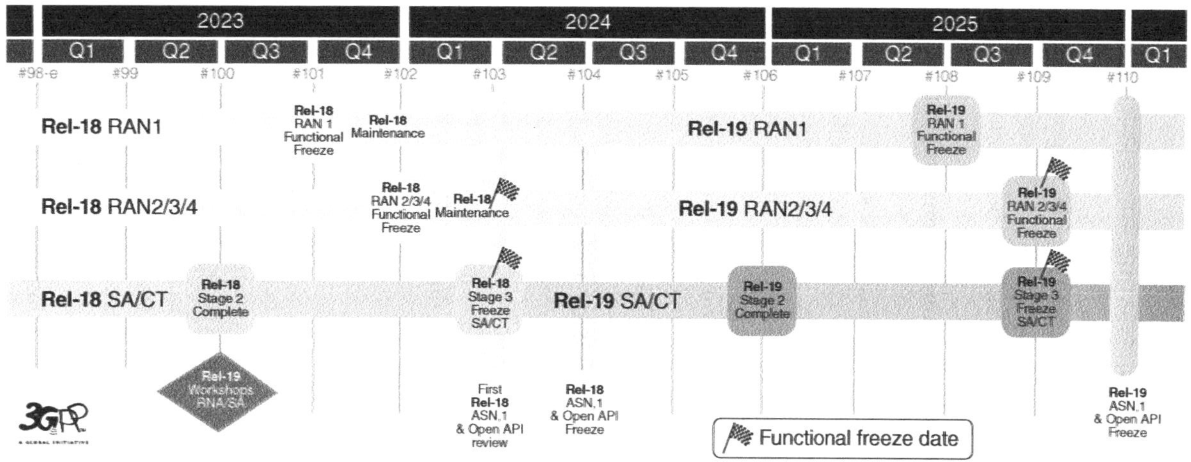

Figure 2.1 The latest 5G timeline from ITU and 3GPP, which uses a system of parallel "releases" providing developers with a stable platform for the implementation of features at any given point. Future dates are proposals and are subject to changes.

2.3 Characteristics of 5G

As in IMT for 2020, there are three requirements for new 5G applications.

(a) **Enhanced Mobile Broadband (eMBB)** defines higher data rates, traffic and connection densities, and user mobility.

This facilitates 5G networks to provide high-speed data and multimedia services such as AR, VR, and high-quality video streaming to mobile users. Technologies such as massive MIMO and beamforming are used to achieve high-speed data transmission.

(b) **Massive Machine-Type Communications (mMTC)** supports very high traffic densities of devices.

In a 5G network, IoT devices such as sensors can talk to each other automatically in large numbers. This is very useful for use cases such as autonomous vehicles and industry automation in which many IoT devices are present. Technologies such as network slicing are used to optimize a use case to achieve high performance.

(c) **Ultra-Reliable Low Latency Communications (URLLC)** provides very low latency and very high communications service availability.

URLLC is very useful for applications using real-time communications, e.g., autonomous vehicles, industrial automation, and VR, typically calling for latency as low as a few milliseconds. To achieve the low latency, technologies such as network slicing for optimizing a specific use case and edge computing by processing data closer to the end user are used.

An illustration of the three requirements is shown in Figure 2.2.

These three requirements also form three use cases for 5G.

2.4 5G New Radio

As discussed earlier, 3GPP issued (in Release 15) a new radio access technology, 5G New Radio (NR), which is based on orthogonal frequency-division multiplexing (OFDM) to support EMBB and MMTC applications through high-speed data and low-latency communications. Succeeding 4G LTE, NR introduces several improvement areas, such as massive MIMO and beamforming technologies. Here we discuss some of these key technologies in NR.

2.4.1 Orthogonal Frequency-Division Multiplexing

OFDM is a signal modulation technique that enhances data rate and spectral efficiency by dividing a main carrier into multiple subcarriers. Based on the

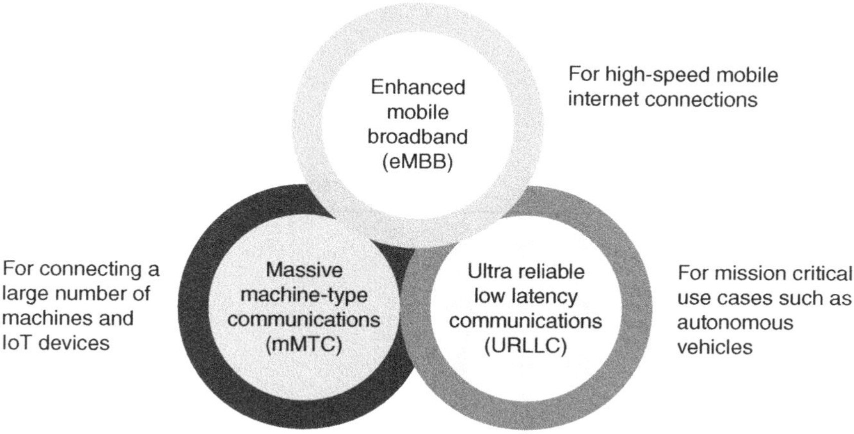

Figure 2.2 5G standard requirements.

frequency-division multiplexing scheme, the subcarrier signals are arranged orthogonally to one another. Figure 2.3 shows different carrier conditions in the frequency domain.

As illustrated in Figure 2.3, a multicarrier arrangement shows that carriers are spaced far apart and, therefore, not overlapping in the frequency domain. In OFDM, adjacent subcarriers are supposed to carry zero-signal magnitude at any frequency when a subcarrier signal is received at its maximum strength by the receiver. This means subcarriers will not interfere with one another and OFDM eliminates crosstalk between the subcarrier signals.

There are advantages of using OFDM over transmission of a single carrier: high data speed is achieved through ODFM since multiple subcarriers are transmitted in parallel simultaneously through a single channel. This also facilitates parallel processing of received subcarrier signals on the receiver. More subcarriers can be added to the channel when there is more data to be transmitted.

In addition, OFDM uses cyclic prefixes to help minimize the multipath fading, which reduces the received signal strength and possibly causes errors. Multipath fading occurs when signals are reflected or diffracted by buildings or other obstacles such as trees, thus arriving at the receiver through different paths and at different times.

The capabilities of transmitting and processing subcarriers in parallel and using the cyclic prefixes to minimize multipath fading effects enable OFDM to reduce latency. Low latency facilitates real-time applications such as video streaming, VR, and AR.

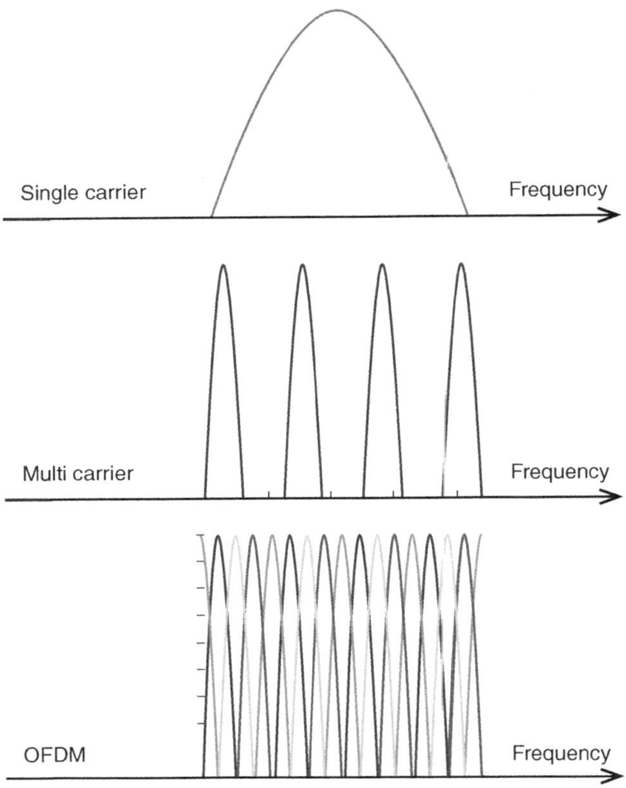

Figure 2.3 Multi-subcarrier signals arrangement in comparison to a single carrier and a multicarrier arrangement. Source: Choudhary and Janyani [6]/with permission of Springer Nature.

2.4.2 Massive MIMO

MIMO involves the use of multiple antennas to transmit and receive wireless signals. This approach has been adopted in 4G LTE. In the simplest form, an MIMO is a 2T2R configuration where two transmitters and two receivers are available. Such a 2×2 MIMO configuration offers relatively low capacity. The illustration of 2×2 and 4×4 MIMO is shown in Figure 2.4. When configured in a 16×16 form or above, e.g., 32×32 or 64×64, it is called a massive MIMO system.

5G makes use of massive MIMO technology in which a high number of antenna elements are present. Due to UE limitations, we use a 4×4 configuration for the receiver. In addition to the high data rate, capacity is also increased through

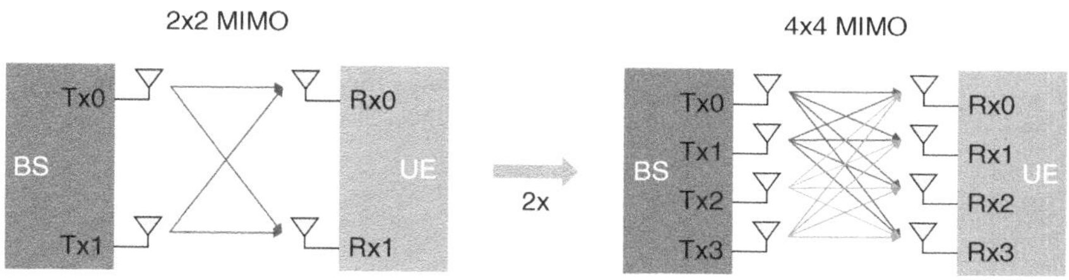

Figure 2.4 Example MIMO configurations: 2 × 2 and 4 × 4. Source: From 5G Indonesia https://www.5g-indonesia.com/2020/08/penjelasan-lengkap-massive-mimo-5g.html.

a technique called multiuser MIMO, rendering such a system is very useful in high-user density areas.

2.4.3 Beamforming

To support massive MIMO, multiple antenna elements are arranged in an array form. The individual antenna elements transmit narrower radio beams and have higher directivity compared to the radio beams from a single antenna configuration. A specific UE device can be targeted this way. Massive MIMO offers advantages over non-massive MIMO systems in terms of coverage due to the fact that more radio beams are available. In addition, since massive MIMO radio beams are narrow, they tend to avoid signals from other transmitters, thus reducing interference. A simplified diagram showing antenna beam patterns from an antenna array is depicted in Figure 2.5. A beamformer controls the phase and directivity of the individual antennas to transmit or receive data effectively to or from a specific device with minimal interference. Beamforming is an important technique used in 5G to tackle interference and limited ranges.

Figure 2.5 Antenna beam targeting particular user equipment. Source: From https://www.electronics-notes.com/articles/connectivity/5g-mobile-wireless-cellular/massive-mimo.php.

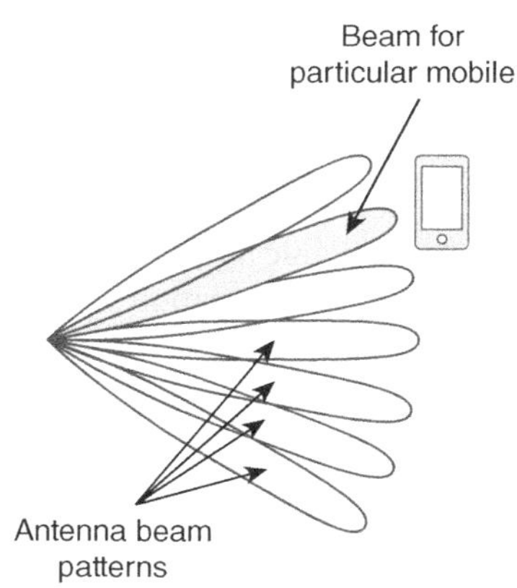

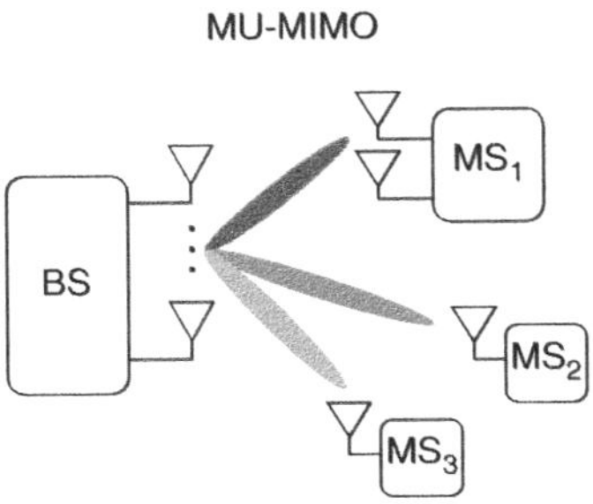

Figure 2.6 Multiuser MIMO configuration. Source: A. Shaikh and M. J. Kaur, 2019/with permission of IEEE. https://www.everythingrf.com/community/what-is-mu-mimo.

2.4.4 Multiuser MIMO

So far, we have discussed how a 5G base station can service a single user. When multiple UEs located in different areas are grouped, the combined resources (in terms of signal frequency or timing) are used for all UEs in the same group. This means that all UEs are served simultaneously, thus improving user capacity. Due to the different UE locations, no interference will occur. A trade-off of using this technique is that UEs close to each other should not be arranged in the same group, as interference may occur. A high separation of the distance between UEs is needed. As shown in Figure 2.6, user MS3 will not interfere with user MS1, as they are not physically close.

2.5 Data Plane and Control Plane Separation

The separation of the control plane and the data plane is a networking and telecommunications concept that involves dividing the tasks and responsibilities of a network device into two distinct components: the control plane and the data plane. These components handle different functions and operate independently to optimize network performance, scalability, and flexibility [7].

Separating the control and data planes is commonly implemented in various network architectures and technologies, including software-defined networking (SDN) and network function virtualization (NFV). These approaches leverage the separation to centralize network control, enhance network programmability, and enable dynamic network management and orchestration.

Here is an explanation of the control plane and the data plane [7]:

(I) Control plane

The control plane is responsible for managing and controlling the overall behavior and operations of the network. It handles tasks related to network configuration, routing protocols, network management, policy enforcement, and control of network devices. The control plane determines how data packets are forwarded, establishes and maintains routing tables, and makes decisions about the most efficient paths for traffic flow.

The control plane also facilitates communication and coordination among network devices by exchanging control messages and updates. For example, in a dynamic routing protocol like Open Shortest Path First (OSPF), the control plane is responsible for exchanging routing information between routers, calculating optimal routes, and updating routing tables accordingly.

(II) Data plane

The data plane, also known as the forwarding plane or forwarding engine, is responsible for the actual forwarding of data packets within the network. It handles the movement of data packets from the source to the destination based on the instructions provided by the control plane. The data plane performs packet switching, forwarding, and filtering functions using predefined rules and policies.

When a data packet arrives at a network device, the data plane examines the packet header, determines the appropriate forwarding path based on forwarding tables or rules, and forwards the packet to the next hop or destination. The data plane operates highly efficiently and deterministically, focusing solely on packet forwarding tasks without involvement in control or decision-making processes.

(III) Advantages of separating the control plane and the data plane

The separation of the control plane and the data plane offers several advantages.

(a) **Scalability**

By separating the control plane from the data plane, network devices can focus on their specific roles more efficiently [7]. This allows for better scalability, as control plane processes can be centralized or distributed across multiple devices while the data plane handles the high-speed forwarding of packets.

(b) **Flexibility and Agility**

With a separate control plane, network administrators can dynamically configure and update network policies, routing protocols, and network settings without disrupting the forwarding of data packets [7]. This separation enables quicker adaptation to changes in network conditions or requirements.

(c) **Fault Isolation**

Separating the control plane and the data plane helps in isolating faults or failures [8]. If there is a problem with the control plane, such as a routing protocol issue, the data plane can continue forwarding packets based on the existing forwarding tables, ensuring uninterrupted network connectivity.

(d) **Security**

By separating the control and data planes, it becomes possible to enforce security measures independently [9]. Access control policies and security mechanisms can be applied at the control plane level to regulate the network's configuration and management, while the data plane can focus on efficient packet forwarding.

2.6 5G Applications

5G technology opens up a wide range of possibilities and enables various applications across different sectors. It is difficult, if not impossible, to achieve such applications before the coming of 5G technology. Figure 2.7 illustrates different applications.

Here are some key applications that are being driven by 5G:

(a) **Enhanced Mobile Broadband (eMBB)**

5G's eMBB capabilities significantly provide higher data speeds, greater capacity, and lower latency compared to previous generations [10, 11]. This enables users to stream high-definition and ultra-high-definition (UHD) video content seamlessly without buffering, even in crowded areas. It also

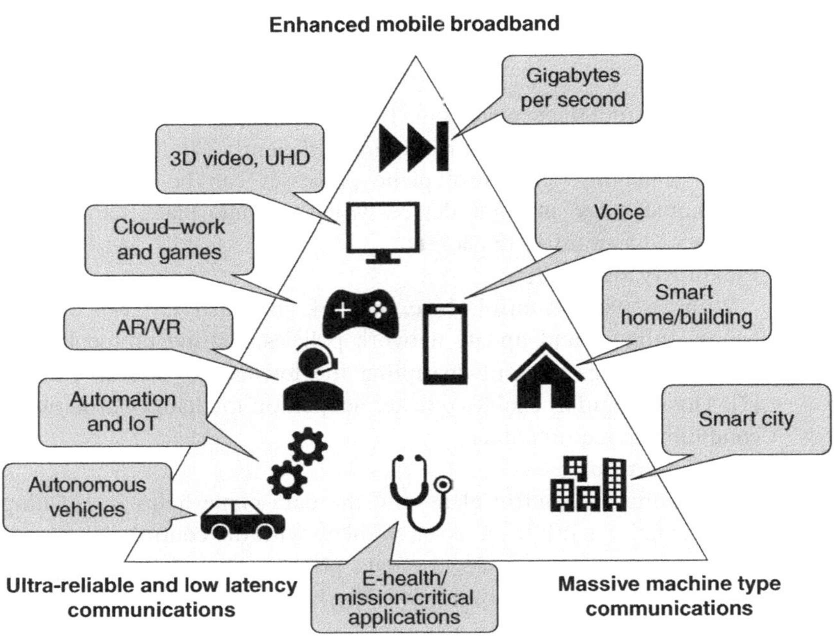

Figure 2.7 Key 5G applications.Source: https://www.commagility.com/applications-and-markets/wireless-networks/5g.

enables faster downloads and uploads, making tasks such as downloading large files, sharing multimedia content, and backing up data quicker and more efficient. 5G's low latency ensures real-time responsiveness, enhancing the user experience for interactive applications, online gaming, and immersive VR/AR experiences.

(b) **Internet of Things (IoT)**

5G is a game changer for the IoT, enabling the connection of a massive number of devices simultaneously. It provides the necessary infrastructure to support the growth of IoT applications across various sectors. 5G's high device density, low power consumption, and extended battery life make it ideal for connecting and managing IoT devices efficiently [12, 13]. Industries such as smart cities, smart homes, industrial automation, agriculture, healthcare, and logistics can benefit from 5G's capabilities for real-time monitoring, data collection, and intelligent decision-making.

(c) **Mission-Critical Communications**

5G's URLLC capabilities are crucial for mission-critical applications that require immediate and dependable connectivity [14, 15]. Industries such as public safety, emergency services, healthcare, and transportation can leverage 5G for real-time communication, video surveillance, and remote operations. In emergencies, 5G enables faster response times, improved coordination, and more effective disaster management [16]. For applications like remote surgeries, autonomous vehicles, and industrial control systems, the low latency of 5G ensures real-time responsiveness, allowing for precise and time-sensitive actions.

(d) **Industrial Automation and Robotics**

5G is a catalyst for the advancement of industrial automation and robotics. It enables real-time communication, low-latency control, and precise coordination of machines, robots, and automated systems [17]. Industries such as manufacturing, logistics, and energy benefit from 5G's capabilities to monitor and control industrial processes remotely, leading to increased productivity, reduced downtime, and enhanced worker safety [18]. With 5G, factories can implement real-time monitoring and predictive maintenance, allowing for proactive repairs and optimization of production processes [19].

(e) **Smart Cities and Infrastructure**

5G plays a pivotal role in the development of smart cities and the transformation of urban infrastructure. It enables the interconnectivity of various components, including sensors, devices, and infrastructure, to improve efficiency, sustainability, and the quality of urban life [20]. Applications like smart grids, intelligent transportation systems, environmental monitoring, and waste management benefit from 5G's fast and reliable connectivity.

With 5G, cities can optimize energy consumption, reduce traffic congestion, enhance public safety, and implement proactive environmental monitoring.

(f) **Healthcare and Telemedicine**

5G revolutionizes healthcare by enabling telemedicine services and remote patient monitoring [21]. It provides the necessary bandwidth and low latency for high-quality video consultations between doctors and patients, regardless of location. Remote patient monitoring using IoT devices, wearables, and connected healthcare devices becomes more feasible and efficient with 5G [22]. Real-time transmission of medical data, including vital signs, enables timely interventions and remote diagnostics, improving patient care, especially in rural or underserved areas.

(g) **Media and Entertainment**

5G offers transformative opportunities in the media and entertainment industry. It enables seamless streaming of high-quality UHD content on mobile devices, eliminating buffering and providing a superior viewing experience. The low latency of 5G enhances interactive gaming experiences, allowing for real-time multiplayer gaming, cloud gaming, and eSports. Immersive technologies like VR and AR benefit from 5G's high bandwidth and low latency, enabling realistic and interactive experiences in areas such as gaming, education, and training [23, 24].

(h) **Precision Agriculture**

5G has a significant impact on the agricultural sector through precision farming techniques. It enables real-time monitoring of soil conditions, weather data, crop health, and machinery, improving resource management and optimizing crop production [25, 26]. 5G connectivity allows for the integration of drones, sensors, and autonomous machinery, enabling precise application of resources, such as water, fertilizers, and pesticides, based on real-time data [27]. This leads to increased crop yield, reduced environmental impact, and improved sustainability in agriculture.

These applications demonstrate the transformative potential of 5G technology across various sectors, driving innovation, efficiency, and new possibilities. As 5G networks continue to expand and evolve, we can expect further advancements and novel use cases to emerge, shaping the way we live, work, and interact in the digital age.

2.7 Smooth Transition to 6G

Field-programmable gate arrays (FPGAs) can significantly facilitate the transition from 5G to 6G in the future. Figure 2.8 shows the Xilinx (AMD) VirtexUltraScale VCU108 FPGA evaluation board, which has high performance, low power

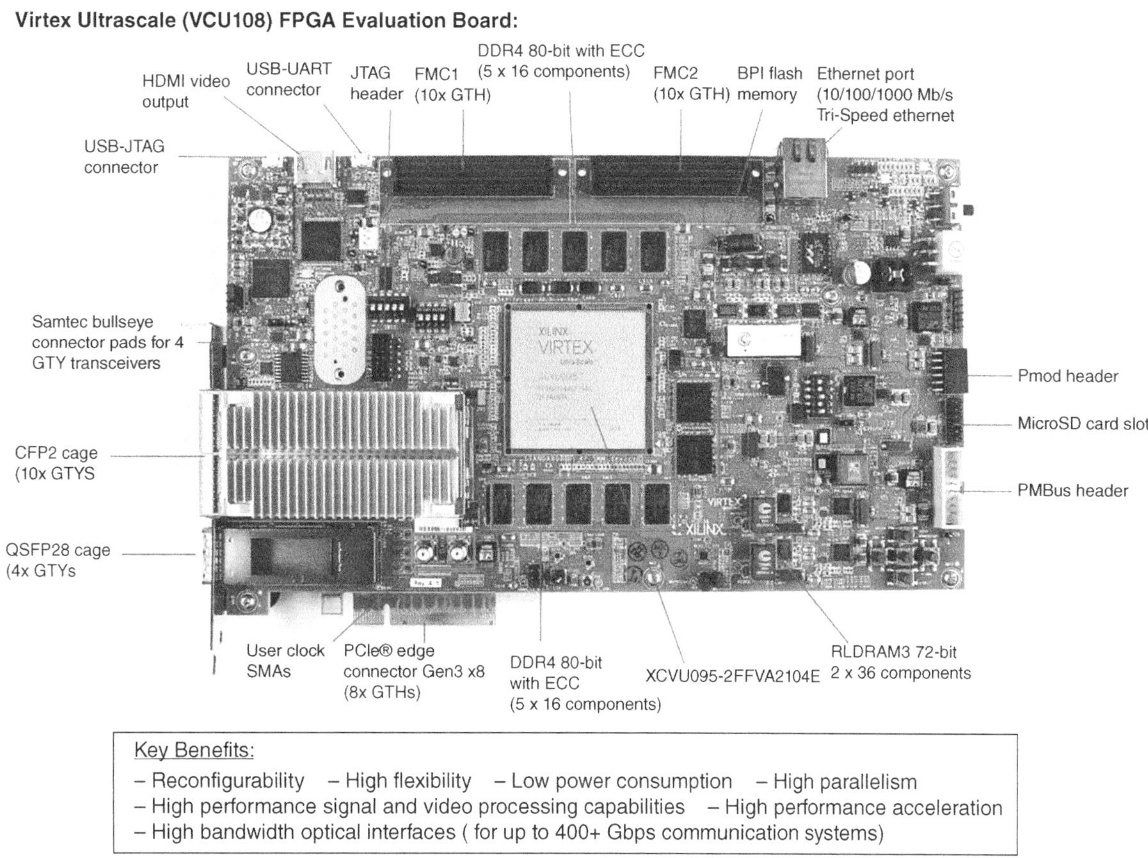

Figure 2.8 AMD Virtex Ultrascale (VCU108) FPGA Evaluation Board. Advanced Micro Devices, Inc. – AMD/https://www.xilinx.com/products/boards-and-kits/ek-u1-vcu108-g.html/Last accessed on June 28, 2024.

consumption, and reconfigurable capabilities. Here are a few ways FPGAs contribute:

(a) **Flexibility and Upgradability**
FPGAs offer inherent flexibility, allowing for the reconfiguration of hardware circuits and functionality [28, 29]. This flexibility is crucial for the transition to 6G because it enables the adoption of new standards, protocols, and features without replacing entire hardware systems.
As 6G evolves, new requirements and technologies will emerge. FPGAs can be reprogrammed to support these changes, ensuring compatibility and seamless integration into existing infrastructure. Upgrading FPGA-based systems to accommodate 6G advancements can be done through software updates, reducing costs and minimizing disruption to network operations.

(b) **Prototyping and Development**
FPGAs serve as an ideal platform for prototyping and development in wireless communication systems [29]. Researchers and developers can utilize FPGAs to implement and test new algorithms, modulation schemes, and network architectures specific to 6G requirements.
FPGA-based prototyping allows for rapid iterations and real-time evaluation of concepts, facilitating the exploration of innovative ideas and accelerating the development of 6G technologies [30].

(c) **High-Speed Data Processing**
6G networks are expected to handle an enormous amount of data at ultra-low latency. FPGAs excel in high-speed data processing due to their parallel processing capabilities and customizable hardware designs [28, 31].
By designing specialized hardware architectures on FPGAs, specific tasks such as complex signal processing, beamforming, and channel estimation can be accelerated, thus meeting the demanding requirements of 6G applications. FPGAs can offload computationally intensive tasks from general-purpose processors, resulting in reduced latency and increased overall system performance.

(d) **Customization and Optimization**
FPGAs offer a high degree of customization, enabling designers to tailor hardware implementations to the unique characteristics and demands of 6G networks. Specific features and functionalities required by 6G, such as massive MIMO, millimeter-wave communications, and advanced modulation schemes, can be implemented efficiently within FPGAs [28].
By optimizing FPGA designs, power consumption, resource utilization, and performance can be balanced, resulting in efficient and optimized 6G

systems. FPGAs also allow for the co-design of hardware and software, enabling system-level optimizations and achieving higher efficiency and performance than off-the-shelf solutions.

(e) **Edge Computing and AI**

6G networks are expected to leverage edge computing capabilities to enable real-time processing and reduced latency for various applications [32]. FPGAs are well-suited for implementing AI algorithms and accelerating AI computations at the network edge [33]. FPGA-based AI accelerators can be integrated into 6G infrastructure, enabling tasks such as real-time data analysis, intelligent resource allocation, network optimization, and intelligent decision-making. By leveraging the parallel processing capabilities of FPGAs, AI computations can be performed with lower latency and higher efficiency, unlocking the potential of AI-driven 6G applications.

(f) **Hardware Security**

With the increasing complexity and security concerns in 6G networks, FPGA-based solutions can enhance hardware security. FPGAs provide the ability to implement security features directly into the hardware design, making it more resistant to attacks [34].

Hardware-level security mechanisms, such as encryption, authentication, and secure key storage, can be integrated into the FPGA fabric, safeguarding critical network functions from potential vulnerabilities. FPGA-based security solutions offer protection against threats like data breaches, unauthorized access, and tampering, ensuring the integrity and confidentiality of 6G networks.

(g) **Advantage Over ASICs**

While ASICs offer benefits such as lower power consumption, higher performance, and potentially lower unit costs for high-volume production, FPGAs excel in terms of flexibility, time-to-market, design iterations, prototyping, and cost-effective development [29]. These advantages make FPGAs preferred when customization, rapid prototyping, flexibility, and shorter development cycles are paramount.

Overall, by harnessing the flexibility, programmability, high-speed data processing capabilities, customization options, and support for edge computing and AI, FPGAs enable a smooth and efficient transition from 5G to 6G. Their versatility and adaptability empower network operators and equipment manufacturers to embrace new technologies, address evolving requirements, and deploy optimized solutions for the next generation of wireless communication.

2.8 6G Expected Timeline, Spectrum, and Characteristics

The development and deployment of 6G technology are still in the early stages, and many aspects are subject to ongoing research and exploration. However, based on industry trends and early discussions, here are some expectations regarding the timeline, spectrum, and characteristics of 6G:

(I) Timeline

The 6G development and deployment timeline is subject to ongoing research and industry collaboration. While the commercial deployment of 6G is anticipated to happen in the 2030s [35], the timeline for various stages of its development can vary. In the early stages, researchers and academia will explore fundamental concepts, conduct feasibility studies, and prototype technologies. This will be followed by standardization efforts led by organizations like the ITU and industry collaborations to define the specifications and requirements of 6G. Afterward, network infrastructure providers and telecommunications companies will begin implementing and deploying 6G networks. The timeline may also depend on factors such as technology maturity, regulatory frameworks, spectrum availability, and market demand.

(II) Spectrum

6G is expected to utilize a combination of existing and new frequency bands to meet its requirements. While specific frequency bands for 6G have not yet been defined, the 6G expected to operate in higher frequency ranges compared to previous generations like 5G. The use of higher frequencies, such as the millimeter-wave (mmWave) spectrum and potentially even terahertz (THz) frequencies, can offer larger bandwidths to accommodate the massive data rates envisaged for 6G [35, 36]. However, higher frequencies pose challenges related to signal propagation and coverage, necessitating advancements in antenna technologies, beamforming, and signal processing techniques to overcome these limitations [37, 38].

(III) Characteristics

Key expected characteristics for 6G are listed below.

(a) **Ultra-High Data Rates**

6G aims to provide significantly higher data rates than 5G, potentially reaching multi-terabit-per-second speeds [39]. This will enable seamless and instantaneous data-intensive applications, such as 8 K/16 K video streaming, VR, AR, and holographic communications. The ultra-high data rates will support massive bandwidth requirements and enable a new wave of immersive and high-resolution experiences.

(b) **Ultra-Low Latency**

6G aims to achieve ultra-low latency, reducing the delay in data transmission to extremely low levels, potentially in the sub-millisecond range [40]. This near-real-time responsiveness is essential for applications that demand immediate interactivity, such as remote surgery, autonomous vehicles, smart manufacturing, and real-time gaming. Ultra-low latency will be enabled by advancements in network architecture, edge computing, distributed processing, and optimized transmission protocols.

(c) **Massive Connectivity**

6G will support a massive number of connected devices, far surpassing the scale of previous generations. It will accommodate the proliferation of IoT devices, sensors, and autonomous systems, creating a truly interconnected world [41]. With seamless connectivity, 6G will enable advanced applications in smart cities, smart homes, industrial automation, and healthcare. This massive connectivity will require efficient network management, optimized resource allocation, and robust security mechanisms.

(d) **Energy Efficiency**

6G will prioritize energy efficiency to address the growing environmental concerns and sustainability requirements. It will aim to optimize power consumption by developing energy-efficient hardware designs, intelligent power management algorithms, and dynamic sleep modes [42]. Advanced technologies like energy harvesting, low-power device architectures, and network optimization techniques will be employed to minimize energy consumption and reduce the carbon footprint of wireless networks.

(e) **Enhanced Security and Privacy**

6G networks will emphasize security and privacy to address the evolving threats and vulnerabilities. It will likely incorporate advanced post-quantum encryption algorithms, authentication mechanisms, and privacy-preserving technologies to ensure secure communication and protect user data [43–46]. Privacy-enhancing techniques, such as differential privacy and secure multi-party computation, will be integrated to give users more control over their personal information while enabling data sharing and analysis.

(f) **Artificial Intelligence Integration**

AI and machine learning will play a crucial role in 6G networks, enabling intelligent network management, optimized resource allocation, and personalized services. AI algorithms will be integrated into network components, enabling autonomous decision-making, real-time network optimization, and predictive analytics [47]. AI will also enable context-aware services, adaptive network configurations, and intelligent traffic routing, enhancing the overall performance and user experience of 6G networks.

(g) **Enhanced Spectrum Efficiency**

6G will focus on maximizing spectrum efficiency to address the increasing demand for wireless connectivity. Advanced techniques such as spectrum sharing, dynamic spectrum access, and cognitive radio will be employed to optimize the use of available frequency bands. Dynamic spectrum management will enable dynamic allocation of resources based on demand, allowing for efficient coexistence of diverse applications and wireless systems. Additionally, beamforming technologies will be utilized to enhance spectral efficiency and improve coverage in higher frequency bands [48].

It is important to note that these expectations are based on the current understanding and research trends in the field of 6G. As the technology evolves and further research is conducted, new characteristics and requirements may emerge, shaping the actual features and capabilities of 6G.

2.9 6G Potential Applications

The potential applications of 6G are still being explored and are subject to ongoing research and development. However, several key areas of interest have emerged, highlighting the envisioned applications of 6G networks, as illustrated in Figure 2.9.

Here are some potential 6G applications:

(a) **Ultra-High-Speed Internet**

6G aims to provide significantly faster data rates compared to 5G, potentially reaching speeds of multiple terabits per second. This ultra-high-speed

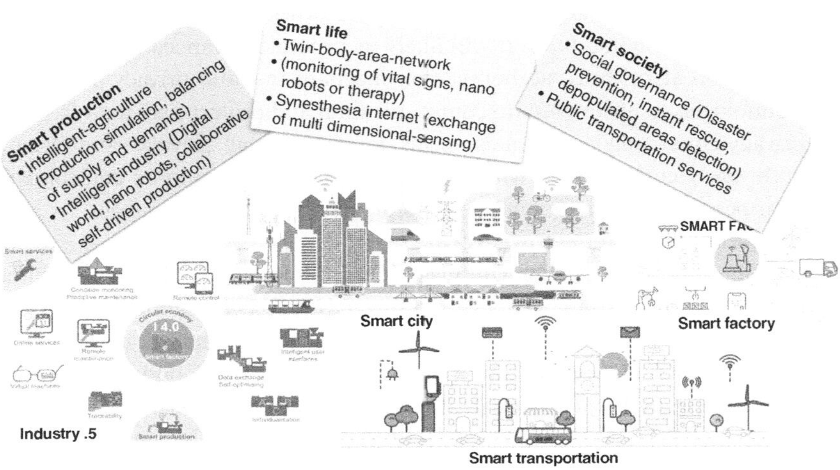

Figure 2.9 Potential 6G applications. Source: Hakeem et al. [49]/with permission of Elsevier.

internet will enable seamless and instantaneous access to content and services, empowering users to stream high-resolution videos, engage in high-quality video conferencing, and download large files in a matter of seconds. It will also support emerging technologies like 8 K and 16 K video streaming, VR, and AR applications, enabling immersive and bandwidth-intensive experiences [50].

(b) **Immersive Extended Reality (XR)**

6G is expected to transform XR technologies by delivering enhanced immersion and realism. With ultra-low latency and high bandwidth, 6G can support real-time rendering and streaming of complex XR environments, making virtual worlds more lifelike and interactive. This will enable users to experience more convincing virtual simulations, realistic telepresence, and seamless integration of digital information into the physical world. It can have applications in gaming, education, training, remote collaboration, and entertainment [51].

(c) **Holographic Communications**

6G envisions the transmission of holographic content, enabling realistic 3D representations of remote individuals or objects in real-time. This technology can revolutionize teleconferencing and communication by providing a more natural and immersive experience. Users will be able to engage in holographic telepresence, where it feels like the person or object is physically present. This can have significant implications for remote collaboration, telemedicine, and virtual tourism [52].

(d) **Internet of Everything (IoE)**

6G aims to create a pervasive and seamless IoE ecosystem, connecting billions of devices and enabling advanced services and applications. It will go beyond the traditional IoT, encompassing a wide range of connected entities, including smart devices, sensors, vehicles, infrastructure, and even biological entities. This will enable intelligent automation, data-driven decision-making, and enhanced connectivity in areas such as smart cities, smart homes, intelligent transportation systems, and precision agriculture [39, 52, 53].

(e) **Autonomous Systems and Robotics**

6G will play a vital role in advancing autonomous systems, including autonomous vehicles, drones, robots, and AI-powered machines. With ultra-low latency and high reliability, 6G networks can support real-time communication and coordination among these systems, enabling safer and more efficient operations. This can lead to advancements in autonomous transportation, delivery services, industrial automation, and smart manufacturing, with applications ranging from self-driving cars to fully automated factories [52, 54].

(f) **Intelligent Infrastructure and Smart Cities**

6G will facilitate the development of intelligent infrastructure and smart city solutions. It will enable seamless connectivity and communication among various elements of a smart city, including smart grids, smart buildings, smart transportation systems, and environmental monitoring systems. This will result in enhanced energy efficiency, optimized resource allocation, improved public safety, and sustainable urban development. For example, 6G can support real-time energy management, intelligent traffic control, and efficient waste management systems [54].

(g) **Personalized Healthcare and Telemedicine**

6G networks will empower personalized healthcare and remote medical services. With ultra-reliable and low-latency connectivity, healthcare professionals can remotely monitor patients, perform remote surgeries, and provide real-time consultations. 6G can enable advanced telemedicine solutions, remote diagnostics, and personalized treatments, ensuring access to healthcare services regardless of geographical limitations. It can also support wearable devices, biosensors, and real-time health monitoring for proactive and preventive healthcare [52, 54, 55].

(h) **Environmental Monitoring and Conservation**

6G can significantly contribute to environmental monitoring, conservation efforts, and sustainability initiatives. By leveraging a vast network of IoT sensors, satellite imagery, and advanced analytics, 6G networks can collect and analyze real-time data related to climate change, air quality, water resources, and ecosystem health [54, 55]. This information can be used to develop efficient environmental management strategies, early warning systems for natural disasters, and sustainable practices in industries such as agriculture, forestry, and energy.

These potential applications are speculative and subject to ongoing research and development. The actual implementation and availability of these applications will depend on technological advancements, regulatory frameworks, business models, and societal needs. As 6G continues to evolve, new possibilities and applications may emerge, shaping the future of connectivity and technology.

2.10 Edge, Fog, and Cloud Computing in Relation to 5G and 6G

Edge computing, fog computing, and cloud computing are three distinct paradigms that can work together to support the capabilities and requirements of 5G and potentially 6G networks. Each of these computing models addresses different aspects of data processing, storage, and communication in the context of modern and future telecommunications systems. A simplified computing infrastructure showing the three technologies is shown in Figure 2.10. This section

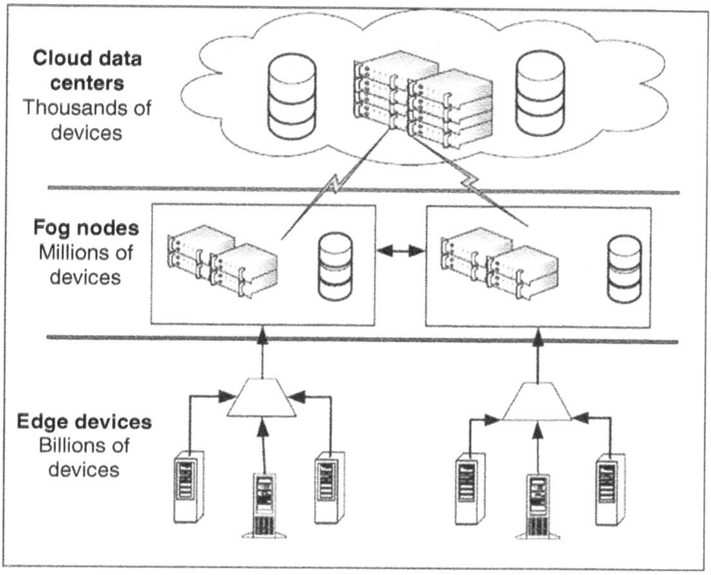

Figure 2.10 Simplified computing infrastructure showing the edge, fog, and cloud [56] Bhat, S.A et al., 2020/with permission of IEEE.

explains the relationship between edge computing, fog computing, cloud computing, and 5G and 6G. We will discuss these three computing technologies one by one.

2.10.1 Edge Computing in Relation to 5G and 6G

Edge computing plays a crucial role in supporting 5G and potentially 6G networks, especially when it comes to processing speed. Here is how edge computing can enhance processing speed in these advanced telecommunication systems:

(a) **Reduced Latency:**
Latency refers to the delay between sending data and receiving a response. Low latency is crucial in applications that demand real-time interaction, like VR or remote robotic control. Edge computing places processing resources closer to where the data is generated, which minimizes the time it takes for the data to travel to a distant cloud server and back. This proximity translates to faster response times, enabling applications to function smoothly without noticeable delays [57, 58].

(b) **Local Data Processing:**
Edge devices, such as IoT sensors or smart cameras, often generate substantial amounts of data. Transmitting all this data to a central cloud for

processing can result in delays due to network congestion and processing times. With edge computing, the devices perform initial data processing, filtering out irrelevant data and sending only the essential parts to the cloud. This local processing accelerates the overall decision-making process, as only the most critical information is sent for further analysis [57, 59].

(c) **Real-Time Applications and Decision Making:**
Certain 5G and 6G network applications require real-time interaction and decision-making, where delays can impact user experience or safety. For example, self-driving cars need to respond instantly to road conditions. Edge computing allows these devices to process data and make decisions locally without waiting for instructions from a distant cloud. This ensures that critical actions are taken swiftly, enhancing safety and efficiency [57, 59].

(d) **Data Filtering and Prioritization:**
In environments with many connected devices, such as smart factories or smart cities, vast amounts of data are generated. Edge computing lets devices prioritize data at the source based on its importance [58]. For instance, a sensor detecting a critical fault might send an alert immediately, while less urgent data can be batched and sent at intervals. This prioritization ensures that vital information reaches its destination quickly.

(e) **Local Caching and Content Delivery:**
Edge servers can cache frequently requested content, such as videos, applications, or updates. When a user requests this content, it is delivered from the nearby edge server instead of a remote cloud data center. This significantly reduces the time it takes to access the content, providing a smoother and faster user experience [59].

(f) **Parallel Processing and Load Balancing:**
Edge computing involves a network of distributed edge devices. These devices can collectively perform processing tasks in parallel, meaning multiple tasks are executed simultaneously. Load balancing ensures that processing tasks are evenly distributed across available resources, maximizing efficiency and speeding up overall processing [56].

(g) **Reduced Bandwidth Usage:**
In 5G and 6G networks, bandwidth is a precious resource. Edge computing alleviates the strain on the network by performing initial data processing at the edge. Instead of sending large volumes of raw data to a central cloud, only relevant or summarized data is transmitted. This reduces the amount of data that needs to be sent over the network, resulting in faster data transfer and less congestion [57].

(h) **Network Efficiency:**
By processing data closer to the source, edge computing reduces the need for data to travel long distances over the network. This reduction in data

transmission contributes to lower network congestion and more efficient bandwidth usage, leading to faster overall data processing and higher network scalability [58].

(i) **Security and Privacy**

Many edge devices are situated on the premises of their owners. Thus, the data are processed and stored locally, alleviating the fear of security and privacy issues in cloud computing. With edge computing, sensitive and private data can be processed and stored locally within a secure environment provided by the enterprises (owners).

(j) **Edge AI and Machine Learning:**

Edge computing often incorporates AI and machine learning capabilities at the edge. This enables devices to make intelligent decisions locally without constant communication with the cloud. For instance, a camera system can identify objects or anomalies on-site, reducing the need to send data to a central server for analysis [57].

In essence, edge computing's emphasis on local processing, quick decision-making, data filtering, and efficient resource utilization aligns well with the requirements of 5G and potentially 6G networks. By leveraging edge computing, these advanced communication systems can achieve the processing speed necessary to support demanding applications and provide seamless user experiences.

2.10.2 Fog Computing in Relation to 5G and 6G

Fog computing, which extends the principles of edge computing to a broader scale, can provide significant support to 5G and 6G networks, especially in terms of processing speed. Here is how fog computing enhances processing speed in these advanced telecommunications systems:

(a) **Edge-to-Fog Hierarchy:**

Fog computing operates in a hierarchical manner, extending the edge computing concept. In this hierarchy, edge devices are the closest to data sources, while fog nodes are situated at various intermediate levels. This architecture enables efficient processing distribution. Edge devices can handle immediate processing, while more complex or resource-intensive tasks can be offloaded to higher-level fog nodes. This hierarchy optimizes processing resources and ensures that tasks are executed as swiftly as possible [60].

(b) **Parallel Processing:**

Fog computing involves a network of interconnected edge devices and fog nodes, creating a distributed environment. This network can handle multiple processing tasks concurrently. This parallel processing capability is crucial for applications that require rapid data analysis and decision-making. In 5G and 6G networks, where vast amounts of data need to be processed quickly, the ability to perform multiple tasks at once significantly enhances overall processing speed [61].

(c) **Optimized Data Processing:**
Fog computing enables data to be processed at various layers of the network hierarchy, from edge devices to fog nodes. This optimization ensures that data is processed at the most appropriate level, considering factors like latency requirements and complexity. For instance, latency-sensitive tasks can be handled at the edge, while more resource-intensive tasks can be offloaded to fog nodes. This dynamic data processing contributes to faster response times [60, 62, 63].

(d) **Low Latency and Real-Time Processing:**
Fog nodes are strategically placed in proximity to the data sources or end-users. This proximity reduces the time it takes for data to travel, leading to lower latency. In applications like remote control of robots or real-time monitoring of critical systems, fog computing's local processing capabilities ensure that data is analyzed and acted upon swiftly, meeting the stringent latency demands of 5G and 6G networks [63].

(e) **Load Balancing and Resource Utilization:**
Fog computing employs load-balancing mechanisms to distribute processing tasks across available fog nodes. This ensures that no single node becomes overwhelmed with tasks, maintaining consistent processing speed. By efficiently utilizing all available resources, fog computing optimizes overall system performance and supports high-speed data processing [62, 63].

(f) **Data Aggregation and Filtering:**
In large-scale networks, various devices generate a significant amount of data. Fog computing enables data aggregation and filtering at the network edge. Instead of sending all raw data to centralized cloud servers, fog nodes can preprocess and filter data, sending only the relevant information for further analysis. This data reduction accelerates the processing pipeline [64].

(g) **Localized Intelligence:**
Fog nodes can host local AI and machine learning models. These models can make decisions based on local data without requiring constant communication with central cloud servers. For example, in a smart city setup, data from IoT sensors such as air quality readings can be processed locally. Communications with central cloud servers are made when abnormal conditions are detected [61].

(h) **Efficient Content Delivery:**
Fog nodes can cache frequently accessed content, applications, or updates. This local caching reduces the need to retrieve data from distant cloud servers,

improving content delivery speed. In scenarios like content streaming or software updates, users experience faster access times due to localized content delivery [65].

(i) **Latency-Aware Task Allocation:**

Fog computing's hierarchical architecture allows for intelligent task allocation based on the latency requirements of different applications. Time-sensitive tasks are directed to appropriate layers for immediate processing, ensuring that critical tasks receive priority and reducing the risk of delays [63].

(j) **Dynamic Scalability:**

Fog computing's distributed nature allows the network to scale dynamically by adding more fog nodes as needed. This scalability ensures that processing speed remains consistent even as the network grows, preventing performance bottlenecks [63].

In summary, fog computing enhances processing speed in 5G and 6G networks through its distributed architecture, optimized data processing, low latency, load balancing, and localized intelligence. By leveraging the capabilities of interconnected edge devices and fog nodes, fog computing supports the real-time demands of modern telecommunications systems, enabling applications to deliver high-speed, low-latency experiences to users.

2.10.3 Cloud Computing in Relation to 5G and 6G

Cloud computing is a key technology that supports 5G and 6G networks, especially in terms of processing speed. Here is how cloud computing can enhance processing speed in these advanced telecommunications systems:

(a) **High Computational Power:**

Cloud data centers are equipped with a vast array of high-performance servers, storage systems, and networking infrastructure. This immense computing power enables them to handle complex and resource-intensive tasks efficiently. Cloud computing can deliver results rapidly for applications requiring substantial processing capabilities, such as large-scale simulations or data-intensive computations [66].

(b) **On-Demand Scalability:**

One of the key advantages of cloud computing is its ability to scale resources up or down based on demand. During peak usage times, such as when many users access a service simultaneously, cloud providers can automatically allocate more resources to ensure that processing speed remains high. This dynamic scalability is crucial for maintaining responsive applications and services [67].

(c) **Distributed Data Centers:**

Major cloud providers operate data centers in various regions across the world. This global distribution enables data to be processed closer to the end-users, reducing the physical distance data needs to travel [68]. As a result, data latency is minimized, leading to faster processing times and improved user experiences, particularly for latency-sensitive applications like real-time communication and gaming.

(d) **Parallel Processing and High-Performance Computing (HPC):**

Cloud computing platforms support parallel processing, where tasks are divided into smaller sub-tasks and processed concurrently on multiple servers [69]. Additionally, cloud providers offer specialized high-performance computing (HPC) instances optimized for computationally intensive workloads [70]. These capabilities significantly accelerate data processing, making them suitable for scientific simulations, genetic research, and other data-intensive tasks.

(e) **AI and Machine Learning Services:**

Cloud providers offer AI and machine learning services that provide pre-trained models, algorithms, and tools. These services enable developers to quickly integrate AI capabilities into their applications without the need for extensive AI expertise. This speed-to-market allows businesses to deploy intelligent applications rapidly, enhancing processing speed for AI-driven tasks [71].

(f) **Real-Time Analytics and Insights:**

Cloud platforms offer tools and services for real-time data analytics, allowing businesses to process and analyze data streams as they arrive [57]. This capability is valuable for applications that require immediate insights, such as monitoring social media sentiment or tracking IoT sensor data for anomalies. Cloud-based real-time analytics enable timely decision-making.

(g) **Content Delivery Networks (CDNs):**

Cloud-based CDNs optimize the delivery of digital content, such as videos, images, and web pages. These CDNs use caching and distribution strategies to store content on servers located closer to end-users. As a result, users experience faster loading times and smoother content access, enhancing the overall processing speed of web-based services [72].

(h) **Cost savings**

Small and medium enterprises use cloud services and infrastructure to avoid the costs of building and installing heavy infrastructures and machines. The cloud provides a more economical means of running their applications with higher reliability, scalability, efficiency, and performance. The enterprises will not be charged when they do not run applications on the cloud. In conclusion, cloud computing's combination of massive computational resources, dynamic scalability, global distribution, parallel processing capabilities, specialized services, and managed automation collectively contribute to enhancing processing speed in 5G and 6G networks. By leveraging cloud resources and services, businesses and developers can accelerate data processing, analysis, and application deployment, meeting the high-speed demands of modern telecommunications systems and applications.

References

1 International Telecommunication Union. About International Telecommunication Union (ITU). *ITU*. https://www.itu.int/en/about/Pages/default.aspx.

2 International Telecommunication Union. IMT Vision – Framework and overall objectives of the future development of IMT for 2020 and beyond. *ITU*. https://www.itu.int/dms_pubrec/itu-r/rec/m/R-REC-M.2083-0-201509-I!g!PDF-E.pdf (accessed on 10 February 2023).

3 3GPP. Introducing 3GPP. *3GPP*. https://3gpp.org/about-us/introducing-3gpp (accessed on 10 February 2023).

4 3GPP (2022). TS38.101–1 NR User Equipment (UE) radio transmission and reception, Part 1: Range 1 Standalone.

5 3GPP (2022). TS38.101–2 NR User Equipment (UE) radio transmission and reception, Part 2: Range 2 Standalone.

6 Choudhary, U. and Janyani, V. (2021). Orthogonal frequency division multiplexing for mitigation of optical fiber channel impairments. *Journal of Optics* 50: 186–200. https://doi.org/10.1007/s12596-021-00697-9.

7 Kreutz, D., Ramos, F.M., Verissimo, P.E. et al. (2014). Software-defined networking: a comprehensive survey. *Proceedings of the IEEE* 103 (1): 14–76.

8 Kumar, G.N., Katsalis, K., Papadimitriou, P. et al. (2021). Failure handling for time-sensitive networks using SDN and source routing. *2021 IEEE 7th International Conference on Network Softwarization (NetSoft)*, 28 June–2 July 2021, Tokyo Japan, pp. 226–234. IEEE.

9 Rahouti, M., Xiong, K., Xin, Y. et al. (2022). SDN security review: threat taxonomy, implications, and open challenges. *IEEE Access* 10: 45820–45854.

10 Khan, B.S., Jangsher, S., Ahmed, A., and Al-Dweik, A. (2022). URLLC and eMBB in 5G industrial IoT: a survey. *IEEE Open Journal of the Communications Society* 3: 1134–1163.

11 Shehab, M.J., Kassem, I., Kutty, A.A. et al. (2021). 5G networks towards smart and sustainable cities: a review of recent developments, applications and future perspectives. *IEEE Access* 10: 2987–3006.

12 Khanh, Q.V., Hoai, N.V., Manh, L.D. et al. (2022). Wireless communication technologies for IoT in 5G: vision, applications, and challenges. *Wireless Communications and Mobile Computing* 2022: 1–12.

13 Li, S., Da Xu, L., and Zhao, S. (2018). 5Ginternet of things: a survey. *Journal of Industrial Information Integration* 10: 1–9.

14 Zhang, Q. and Fitzek, F.H. (2015). Mission critical IoT communication in 5G. *Future Access Enablers for Ubiquitous and Intelligent Infrastructures: First International Conference, FABULOUS 2015*, Ohrid, Republic of Macedonia (23–25 September 2015). *Revised Selected Papers 1*, 35–41. Springer International Publishing.

15 Mohammed, N.A., Mansoor, A.M., and Ahmad, R.B. (2019). Mission-critical machine-type communication: an overview and perspectives towards 5G. *IEEE Access* 7: 127198–127216.

16 Spada, M.R., Pérez-Romero, J., Sanchoyerto, A. et al. (2019). Management of mission critical public safety applications: the 5G ESSENCE project. *2019 European Conference on Networks and Communications (EuCNC)*, 18–21 June 2019, Valencia Spain, pp. 155–160. IEEE.

17 Brown, G., Analyst, P., and Reading, H. (2018). Ultra-reliable low-latency 5G for industrial automation. Heavy Reading White Paper on behalf of Qualcomm.

18 Ansari, J., Andersson, C., de Bruin, P. et al. (2022). Performance of 5g trials for industrial automation. *Electronics* 11 (3): 412.

19 Shi, Y., Han, Q., Shen, W., and Zhang, H. (2019). Potential applications of 5G communication technologies in collaborative intelligent manufacturing. *IET Collaborative Intelligent Manufacturing* 1 (4): 109–116.

20 Rao, S.K. and Prasad, R. (2018). Impact of 5G technologies on smart city implementation. *Wireless Personal Communications* 100: 161–176.

21 Liu, E., Effiok, E., and Hitchcock, J. (2020). Survey on health care applications in 5G networks. *IET Communications* 14 (7): 1073–1080.

22 Wang, Y., Tran, P., and Wojtusiak, J. (2022). From wearable device to OpenEMR: 5G edge centered telemedicine and decision support system. *HEALTHINF*, 9–11 February 2022, Virtual Conference, pp. 491–498.

23 Lema, M.A., Laya, A., Mahmoodi, T. et al. (2017). Business case and technology analysis for 5G low latency applications. *IEEE Access* 5: 5917–5935.

24 Konstantoudakis, K., Breitgand, D., Doumanoglou, A. et al. (2022, 2022). Serverless streaming for emerging media: towards 5G network-driven cost optimization: a real-time adaptive streaming FaaS service for small-session-oriented immersive media. *Multimedia Tools and Applications* 81: 12211–12250.

25 Arrubla-Hoyos, W., Ojeda-Beltrán, A., Solano-Barliza, A. et al. (2022). Precision agriculture and sensor systems applications in Colombia through 5G networks. *Sensors* 22 (19): 7295.

26 Naqvi, S.M.Z.A., Saleem, S.R., Tahir, M.N. et al. (2022). Role of 5G and 6G technology in precision agriculture. *Environmental Sciences Proceedings* 23 (1): 3.

27 Shafi, U., Mumtaz, R., García-Nieto, J. et al. (2019). Precision agriculture techniques and practices: from considerations to applications. *Sensors* 19 (17): 3796.

28 Visconti, P., Velazquez, R., Soto, C.D.V., and De Fazio, R. (2021). FPGA based technical solutions for high throughput data processing and encryption for 5G communication: a review. *TELKOMNIKA (Telecommunication Computing Electronics and Control)* 19 (4): 1291–1306.

29 Chamola, V., Patra, S., Kumar, N., and Guizani, M. (2020). FPGA for 5G: re-configurable hardware for next generation communication. *IEEE Wireless Communications* 27 (3): 140–147.

30 Jeon, H.B., Kim, S.M., Moon, H.J. et al. (2023). Free-space optical communications for 6G wireless networks: challenges, opportunities, and prototype validation. *IEEE Communications Magazine* 61 (4): 116–121.

31 Ricart-Sanchez, R., Malagon, P., Salva-Garcia, P. et al. (2018). Towards an FPGA-accelerated programmable data path for edge-to-core communications in 5G networks. *Journal of Network and Computer Applications* 124: 80–93.

32 Riggio, R., Coronado, E., Linder, N. et al. (2021). AI@ EDGE: a secure and reusable artificial intelligence platform for edge computing. *2021 Joint European Conference on Networks and Communications and 6G Summit (EuCNC/6G Summit)*, 8–11 June 2021, Porto Portugal, pp. 610–615. IEEE.

33 Letaief, K.B., Shi, Y., Lu, J., and Lu, J. (2021). Edge artificial intelligence for 6G: vision, enabling technologies, and applications. *IEEE Journal on Selected Areas in Communications* 40 (1): 5–36.

34 Messaoud, S., Amdouni, R., Albouchi, A. et al. (2023). Future internet of things: connecting the unconnected world and things based on 5/6G networks and embedded technologies. In: *Internet of Things-New Trends, Challenges and Hurdles*, 11–12. IntechOpen.

35 Salameh, A.I. and El Tarhuni, M. (2022). From 5G to 6G – challenges, technologies, and applications. *Future Internet* 14 (4): 117.

36 Rappaport, T.S., Xing, Y., Kanhere, O. et al. (2019). Wireless communications and applications above 100 GHz: opportunities and challenges for 6G and beyond. *IEEE Access* 7: 78729–78757.

37 Tripathi, S., Sabu, N.V., Gupta, A.K. and Dhillon, H.S. (2021). Millimeter-wave and terahertz spectrum for 6G wireless. In *6G Mobile Wireless Networks* (pp. 83–121). Cham: Springer International Publishing Yulei Wu, Sukhdeep Singh,

Tarik Taleb, Abhishek Roy, Harpreet S. Dhillon, Madhan Raj Kanagarathinam, Aloknath De.

38 Shafie, A., Yang, N., Han, C. et al. (2022). Terahertz communications for 6G and beyond wireless networks: challenges, key advancements, and opportunities. *IEEE Network* 37 (3): 162–169.

39 Chowdhury, M.Z., Shahjalal, M., Ahmed, S., and Jang, Y.M. (2020). 6G wireless communication systems: applications, requirements, technologies, challenges, and research directions. *IEEE Open Journal of the Communications Society* 1: 957–975.

40 Berardinelli, G., Mahmood, N.H., Rodriguez, I., and Mogensen, P. (2018). Beyond 5G wireless IRT for industry 4.0: design principles and spectrum aspects. In *2018 IEEE Globecom Workshops (GC Workshops)*, 9–13 December 2018, Abu Dhabi UAE, pp. 1–6. IEEE.

41 Katz, M., Matinmikko-Blue, M., and Latva-Aho, M. (2018). 6 Genesis flagship program: building the bridges towards 6G-enabled wireless smart society and ecosystem. *2018 IEEE 10th Latin-American Conference on Communications (LATINCOM)*, 14–16 November 2018, Guadalajara Mexico, pp. 1–9. IEEE.

42 Shen, S., Yu, C., Zhang, K. et al. (2021). Adaptive and dynamic security in AI-empowered 6G: from an energy efficiency perspective. *IEEE Communications Standards Magazine* 5 (3): 80–88.

43 Ziegler, V., Schneider, P., Viswanathan, H. et al. (2021). Security and trust in the 6G era. *IEEE Access* 9: 142314–142327.

44 Yerrapragada, A.K., Eisman, T., and Kelley, B. (2021). Physical layer security for beyond 5G: ultra secure low latency communications. *IEEE Open Journal of the Communications Society* 2: 2232–2242.

45 Porambage, P., Gür, G., Osorio, D.P.M. et al. (2021). The roadmap to 6G security and privacy. *IEEE Open Journal of the Communications Society* 2: 1094–1122.

46 Wang, M., Zhu, T., Zhang, T. et al. (2020). Security and privacy in 6G networks: new areas and new challenges. *Digital Communications and Networks* 6 (3): 281–291.

47 Shehzad, M.K., Rose, L., Butt, M.M. et al. (2022). Artificial intelligence for 6G networks: technology advancement and standardization. *IEEE Vehicular Technology Magazine* 17 (3): 16–25.

48 Hong, W., Jiang, Z.H., Yu, C. et al. (2021). The role of millimeter-wave technologies in 5G/6G wireless communications. *IEEE Journal of Microwaves* 1 (1): 101–122.

49 Hakeem, S.A.A., Hussein, H.H., and Kim, H. (2022). Vision and research directions of 6G technologies and applications. *Journal of King Saud University-Computer and Information Sciences* 34 (6): 2419–2442.

50 You, X., Wang, C.X., Huang, J. et al. (2021). Towards 6G wireless communication networks: vision, enabling technologies, and new paradigm shifts. *Science China Information Sciences* 64: 1–74.

51 De Alwis, C., Kalla, A., Pham, Q.V. et al. (2021). Survey on 6G frontiers: trends, applications, requirements, technologies and future research. *IEEE Open Journal of the Communications Society* 2: 836–886.

52 Nayak, S. and Patgiri, R. (2022). 6G communication: a vision on the potential applications. *Edge Analytics: Select Proceedings of 26th International Conference—ADCOM 2020*, 26–28 August 2020, Singapore, pp. 203–218. Springer Singapore.

53 Alsabah, M., Naser, M.A., Mahmmod, B.M. et al. (2021). 6G wireless communications networks: a comprehensive survey. *IEEE Access* 9: 148191–148243.

54 Asghar, M.Z., Memon, S.A., and Hämäläinen, J. (2022). Evolution of wireless communication to 6G: potential applications and research directions. *Sustainability* 14 (10): 6356.

55 Hewa, T., Gür, G., Kalla, A. et al. (2020). The role of blockchain in 6G: Challenges, opportunities and research directions. *2020 2nd 6G Wireless Summit (6G SUMMIT)*, 17–20 March 2020, Levi Finland, pp. 1–5.

56 Bhat, S.A., Sofi, I.B., and Chi, C.Y. (2020). Edge computing and its convergence with blockchain in 5G and beyond: security, challenges, and opportunities. *IEEE Access* 8: 205340–205373.

57 Dogra, A., Jha, R.K., and Jain, S. (2020). A survey on beyond 5G network with the advent of 6G: architecture and emerging technologies. *IEEE Access* 9: 67512–67547.

58 Hassan, N., Yau, K.L.A., and Wu, C. (2019). Edge computing in 5G: a review. *IEEE Access* 7: 127276–127289.

59 Pham, Q.V., Fang, F., Ha, V.N. et al. (2020). A survey of multi-access edge computing in 5G and beyond: fundamentals, technology integration, and state-of-the-art. *IEEE Access* 8: 116974–117017.

60 Bermbach, D., Pallas, F., Pérez, D.G. et al. (2018). A research perspective on fog computing. *Service-Oriented Computing–ICSOC 2017 Workshops: ASOCA, ISyCC, WESOACS, and Satellite Events*, Málaga, Spain (13–16 November 2017). Revised Selected Papers, 198–210. Springer International Publishing.

61 Santos, J., Wauters, T., Volckaert, B., and De Turck, F. (2017). Fog computing: enabling the management and orchestration of smart city applications in 5G networks. *Entropy* 20 (1): 4.

62 Khakimov, A., Muthanna, A., and Muthanna, M.S.A. (2018). Study of fog computing structure.*2018 IEEE Conference of Russian Young Researchers in Electrical and Electronic Engineering (EIConRus)*, 29 January–1 February 2018, Petersburg Russia, 51–54. IEEE.

63 Meng, Y., Naeem, M.A., Almagrabi, A.O. et al. (2020). Advancing the state of the fog computing to enable 5G network technologies. *Sensors* 20 (6): 1754.

64 Parygin, D.S., Finogeev, A.G., Kamaev, V.A. et al. (2017). A convergent model for distributed processing of big sensor data in urban engineering networks. *Journal of Physics: Conference Series* 803 (1): 012112.

65 Yang, P., Zhang, N., Bi, Y. et al. (2017). Catalyzing cloud-fog interoperation in 5G wireless networks: an SDN approach. *IEEE Network* 31 (5): 14–20.

66 Rommel, S., Raddo, T.R., and Monroy, I.T. (2018). Data center connectivity by 6G wireless systems. *2018 Photonics in Switching and Computing (PSC)*, 19–21 September 2018, Limassol Cyprus, pp. 1–3. IEEE.

67 Wübben, D., Rost, P., Bartelt, J. et al. (2014). Benefits and impact of cloud computing on 5G signal processing. *IEEE Signal Processing Magazine* 31 (6): 35–44.

68 Salhab, N., Rahim, R., and Langar, R. (2020). Optimization of virtualization cost, processing power and network load of 5G software-defined data centers. *IEEE Transactions on Network and Service Management* 17 (3): 1542–1553.

69 Feng, C., Yu, K., Bashir, A.K. et al. (2021). Efficient and secure data sharing for 5G flying drones: a blockchain-enabled approach. *IEEE Network* 35 (1): 130–137.

70 Mangiatordi, F. and Pallotti, E. (2021). A GPU accelerated framework for monitoring LTE/5G interference to DVB-T systems. *2021 AEIT International Annual Conference (AEIT)*, 4–8 October 2021, Virtual Conference, pp. 1–6. IEEE.

71 Boudi, A., Bagaa, M., Pöyhönen, P. et al. (2021). AI-based resource management in beyond 5G cloud native environment. *IEEE Network* 35 (2): 128–135.

72 Alvarez, F., Breitgand, D., Griffin, D. et al. (2019). An edge-to-cloud virtualized multimedia service platform for 5G networks. *IEEE Transactions on Broadcasting* 65 (2): 369–380.

Part II

Theory

3

High-Level Synthesis (HLS)

High-level synthesis (HLS) is a process in digital design and electronic system design where a high-level description of a desired digital circuit or system is automatically transformed into a register-transfer level (RTL) implementation. As summarized in Ref. [1], HLS can be defined as: *The process of converting an untimed or partially timed behavioral description into efficient hardware.* It involves the synthesis of a hardware design from a higher-level, abstract representation, such as a behavioral description written in a high-level programming language like C, C++, or SystemC.

The main idea behind HLS is to allow designers to express their designs in a more abstract and algorithmic manner, leveraging their knowledge and expertise in software programming rather than dealing with low-level hardware details [2, 3]. By using HLS, designers can focus on the functional behavior and performance of the design, while leaving the details of the actual hardware implementation to the synthesis tools. Figure 3.1 illustrates an overview of HLS.

The HLS process flow [4] typically involves the following steps:

1) **Behavioral Modeling:** The designer describes the intended functionality of the circuit or system using a high-level programming language. This description may include algorithms, control flow, data types, and timing constraints.
2) **Compilation:** The high-level description is fed into an HLS tool, which analyzes the code and performs various transformations and optimizations. These transformations include scheduling, resource allocation, binding, and pipelining, among others.
3) **RTL Generation:** Based on the synthesized behavioral model, the HLS tool generates the RTL representation of the design. RTL describes the digital circuit at a lower level of abstraction, capturing the structure, signals, and operations of the hardware components.
4) **Verification and Analysis:** The generated RTL code is verified to ensure its correctness and functionality. Various analysis techniques, such as formal

Edge Computing Acceleration: From 5G to 6G and Beyond, First Edition.
Patrick Hung, Hongwei Kan, and Greg Knopf.

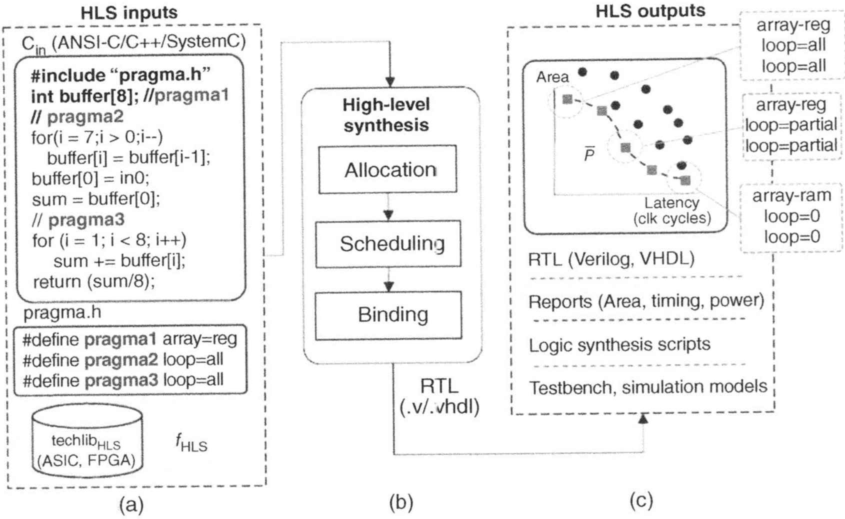

Figure 3.1 Overview of HLS showing (a) input, (b) key tasks, and (c) output. Source: [1] Schaefer, B.C., 2023/with permission from Benjamin Carrion Schaefer.

verification, simulation, and timing analysis, may be applied to validate the design.

5) **RTL Synthesis:** Once the RTL design is verified, it can be further synthesized using traditional RTL synthesis tools to generate a gate-level netlist. This netlist represents the design using logic gates, flip-flops, and other physical components of the target technology.

HLS allows for faster design iterations, as changes can be made at a higher level of abstraction. It enables design exploration and optimization by providing the designer with more control over the architecture and resource allocation. HLS also facilitates the reuse of intellectual property (IP) cores and promotes the adoption of system-level design methodologies.

However, it is worth noting that HLS is not a magic solution and may not always produce optimal results. The quality of the generated RTL code heavily depends on the quality of the initial high-level description and the expertise of the designer in guiding the synthesis process. Nevertheless, HLS has become an increasingly important technology in modern electronic design automation (EDA), enabling faster and more productive hardware design flows.

Currently, there are many more software programmers than hardware programmers in every country in the world. One key advantage of HLS is that it empowers software programmers to design hardware. Software programmers write in high-level programming languages, such as Java, C, C++, or Python, whereas hardware programmers write in hardware description languages (HDLs), such as Verilog, VHDL, or SystemVerilog.

Furthermore, hardware programming needs to address the inherent characteristics of hardware systems, which require special attention to areas of concurrency, reactivity, timing, and clocking [5]. Failure to handle these areas properly can result in incorrect behavior, system instability, timing violations, or even hardware failures. However, it is not typical for software programming to address these areas. Traditionally, hardware programmers work in IC design. With the recent trend of hardware acceleration, there is an increasing demand for hardware programming where HLS can assist. The next subsection will look into the reasons for using HLS.

3.1 Why Use High-Level Synthesis?

There are several reasons why HLS is beneficial and why it is increasingly adopted in the design of complex digital systems [1, 6, 7]. Here are some key advantages of using HLS:

1) **Productivity and Design Exploration:** HLS raises the level of abstraction by allowing designers to express their designs using high-level programming languages such as C, C++, or SystemC. This abstraction level enables designers to focus on the algorithmic aspects of the system rather than low-level hardware details. Using familiar programming languages allows designers to leverage their existing software development skills, resulting in increased productivity. They can quickly prototype and iterate their designs, exploring different algorithms and design alternatives in a shorter time frame.
2) **Faster Time-to-Market:** HLS significantly reduces the design cycle by automating the process of converting high-level descriptions into hardware implementations. Manual RTL coding is time-consuming and error-prone, but with HLS, designers can automatically generate the hardware design from the high-level description. This automation streamlines the design flow, accelerates the development process, and ultimately shortens the time-to-market for the final product.
3) **Improved Design Quality:** HLS tools perform various optimizations during the synthesis process. These optimizations include resource allocation, scheduling, memory optimization, and more. By automatically applying these optimizations, HLS tools can improve the design quality in terms of performance, power consumption, and area utilization. Additionally, HLS tools offer built-in error checking and verification capabilities, which help identify potential issues and improve the correctness of the design.
4) **Design Portability and Reusability:** High-level descriptions written in languages like C or C++ are portable across different platforms and can be reused

for different target architectures. This portability allows designers to explore different hardware platforms and make architectural choices based on performance, power, or other requirements without rewriting the entire design. HLS promotes modular design practices, enabling designers to encapsulate functionality in reusable IP blocks. These blocks can be easily integrated into future designs, saving time and effort.

5) **Design Space Exploration:** HLS provides an efficient way to explore and analyze the design space. Designers can easily experiment with different algorithms, architectures, and optimization strategies. HLS tools offer performance estimation and profiling capabilities, allowing designers to evaluate the impact of different design choices on metrics such as latency, throughput, power consumption, and area utilization. This exploration facilitates finding the optimal trade-offs and making informed design decisions.
6) **Better Abstraction and Refinement:** HLS allows designers to start with a high-level specification and gradually refine the design by adding details and constraints at lower abstraction levels. Designers can incrementally add hardware-specific optimizations and constraints as needed. This iterative refinement process helps manage design complexity and ensures that the design meets the desired requirements. It also provides a clear path for refinement from the high-level description to the low-level hardware implementation.
7) **Bridging the Gap Between Software and Hardware:** HLS bridges the gap between software and hardware design domains. It allows software designers to express their algorithms in a high-level language and automatically generate hardware implementations without requiring extensive knowledge of hardware design. This promotes collaboration between software and hardware teams and facilitates the co-design of complex systems. The software designers can focus on algorithmic development, while the hardware designers can focus on optimizing the hardware implementation.

3.1.1 Hardware Acceleration with High-Level Synthesis

In addition to the reasons above, Schaefer [1] provides an analysis of how HLS is beneficial to hardware accelerators.

HLS is important in modern computer architecture because most integrated circuits (ICs) now consist of heterogeneous system-on-chip (SoC). These SoCs have different components like processors, memory, peripherals, and dedicated hardware accelerators that are all connected. HLS is especially useful for designing these hardware accelerators.

HLS is a good choice for designing accelerators because the applications that need to be accelerated are often initially developed using high-level languages like

ANSI-C, C++, or MATLAB. These applications include things like encryption algorithms, digital signal processing, image processing, and artificial neural networks. HLS allows these applications to be directly transformed into dedicated hardware accelerators, which makes them faster and more energy-efficient by taking advantage of parallel processing.

Compared to general-purpose CPUs and GPUs, dedicated hardware accelerators provide better performance per watt. CPUs can handle a wide range of applications effectively, while GPUs are good at specific tasks like image processing and AI. However, dedicated hardware accelerators offer superior performance per watt for a single application, and that is why they are becoming more common in modern SoCs.

Designing heterogeneous SoCs using low-level HDLs is slow and prone to errors because the design and verification process is complex. Therefore, new design methodologies like HLS are needed to improve productivity. HLS has been widely adopted since the mid-2000s, particularly for designing hardware accelerators within ICs.

In the past, IC design methodologies evolved from manual transistor drawing to connecting gates using HDLs like Verilog and VHDL, which were then transformed into gate netlists using logic synthesis tools. HLS has become the preferred approach for designing ICs, especially for hardware accelerators, because it is more efficient and increases productivity.

Overall, HLS offers numerous advantages, including increased productivity, faster time-to-market, improved design quality, design portability and reusability, efficient design space exploration, better abstraction and refinement, and bridging the gap between software and hardware design domains. These benefits make HLS a valuable tool for designing complex digital systems.

3.2 Common HLS Languages and Platforms

The common HLS languages include C/C++, SystemC, and MATLAB/Simulink [7, 8]. While C/C++ is widely used and provides a rich feature set, SystemC offers a hardware description and verification framework. MATLAB/Simulink is popular for signal processing and control systems design. From the platform side, Intel FPGA [9] and Xilinx FPGA [7, 10] are prominent options. Intel FPGA provides tools like Intel HLS and Quartus Prime, while Xilinx offers Vivado HLS. These platforms support HLS from C/C++ and provide optimizations and integration options. Additionally, Mentor Graphics Catapult [11] and Cadence Stratus HLS [12] are popular HLS tools with C/C++ input support, advanced optimizations, and integration with simulation tools. LegUp is an open-source HLS tool that transforms C/C++ into synthesizable RTL for FPGAs, leveraging LLVM compiler infrastructure [13]. Other HLS tools include the Catapult HLS, Cadence Stratus HLS, and Nymble HLS.

When choosing an HLS language or platform, factors such as tool maturity, target platform compatibility, optimization capabilities, and integration options should be considered. The selection depends on the specific project requirements and the target FPGA vendor. By leveraging these HLS languages and platforms, developers can efficiently generate hardware designs from high-level code, enabling faster development and optimization of FPGA-based systems.

In this chapter, we will discuss, in particular, the following HLS languages and platforms: CUDA, OpenCL, and Maxeler MaxJ.

3.2.1 Compute Unified Device Architecture (CUDA)

CUDA is a programming model and platform developed by NVIDIA for parallel computing on NVIDIA GPUs [14]. While CUDA is not traditionally considered an HLS language, it can be utilized in certain contexts as a higher-level abstraction for hardware design. CUDA simplifies the process of harnessing the power of graphics processing units (GPUs) for general-purpose computing. Using CUDA, programmers can write parallel code using familiar programming languages like C and C++, which can then be executed on GPUs [15].

Developers can leverage the GPUs' massive parallel processing with CUDA, by dividing workloads into thousands of parallel threads [16]. CUDA provides a hierarchical memory model and explicit memory management functions, enabling efficient data movement between the CPU and GPU memory. Additionally, CUDA offers a rich set of libraries optimized for GPU computing tasks, along with a suite of development tools for debugging and performance analysis. Overall, CUDA allows programmers to focus on the algorithmic aspects of their applications while taking advantage of GPU parallelism for improved performance. The following section examines how CUDA can be combined with HLS tools for hardware acceleration.

3.2.1.1 CUDA and HLS for Hardware Acceleration

By combining CUDA with HLS tools, one can explore the potential of hardware acceleration beyond GPUs [17–19]. Working directly with GPUs through CUDA allows one to leverage the existing parallel processing capabilities of GPUs and achieve significant performance gains without the need for additional hardware synthesis and implementation steps. Here is a step-by-step breakdown of the process:

1) **Algorithm Development:** Start by developing one's algorithm using a high-level programming language like C or C++. Identify the computationally intensive portions of the algorithm that could benefit from parallel execution on a GPU.

2) **CUDA Implementation:** Utilize CUDA to implement GPU kernels for the identified compute-intensive parts of the algorithm. CUDA provides a programming model and extensions to C/C++ that allow to express parallel computations explicitly. By writing CUDA kernels, one can take advantage of the parallel processing capabilities of GPUs.
3) **GPU Integration:** GPUs are designed to handle parallel workloads efficiently. With CUDA, one can seamlessly integrate GPU-accelerated kernels into the existing codebase, leveraging the computational power of GPUs without the need for HLS tools at this stage.
4) **Hardware Acceleration Evaluation:** Evaluate the performance of the algorithm on the GPU. Measure factors like execution time, throughput, and speedup achieved compared to running the algorithm solely on the CPU. This step helps to understand the potential benefits of GPU acceleration.
5) **HLS Integration (Optional):** If one requires further acceleration or wants to target specific hardware like FPGAs or ASICs, one can integrate CUDA with HLS tools. These tools analyze the CUDA code and automatically generate RTL designs, which describe the behavior of hardware components needed for acceleration.
6) **Hardware Synthesis and Implementation:** Use the HLS tools to perform hardware synthesis, converting the RTL descriptions into gate-level netlists suitable for hardware implementation. The synthesized design can then be mapped onto the target hardware platform, such as FPGAs, using vendor-specific tools. Verification is performed to ensure the correctness and functionality of the hardware design.

3.2.1.2 Advantage of Using CUDA and HLS for Hardware Acceleration

Using CUDA with HLS tools provides numerous advantages, collectively enabling developers to achieve efficient hardware acceleration, explore various hardware options, and deliver high-performance solutions more efficiently and timely. The following is a summary of these benefits based on Refs. [14, 16–19].

1) **Exploiting GPU Parallelism:** CUDA enables to harness of the immense parallel processing power of GPUs. GPUs are designed with a large number of cores that can perform computations simultaneously. By expressing compute-intensive parts of the algorithm as CUDA kernels, one can take advantage of parallel execution on the GPU. This parallelism allows for the efficient processing of large amounts of data and can lead to significant performance gains compared to running the same code on a CPU.
2) **Higher Productivity and Abstraction:** CUDA provides a higher level of programming abstraction compared to low-level HDLs used in HLS. It allows one to express parallel algorithms using familiar C/C++ syntax, making writing and understanding parallel code easier. CUDA abstracts away many low-level

details of GPU programming, such as thread management, memory management, and synchronization, allowing developers to focus on the algorithm's logic rather than the intricate details of hardware design. This higher level of abstraction increases productivity and reduces the learning curve associated with hardware acceleration.

3) **Utilizing Existing GPU Infrastructure:** GPUs are widely available and commonly used in various computing systems, ranging from personal computers to data centers. By leveraging CUDA for hardware acceleration, one can tap into the existing GPU infrastructure without the need for additional hardware implementations. This advantage allows one to use the substantial computational power already present in GPUs, saving time and resources that would otherwise be spent on developing custom hardware.
4) **Rapid Prototyping and Iteration:** CUDA's high-level programming model and compatibility with GPUs enable rapid prototyping and iteration. One can quickly develop and test GPU-accelerated versions of the algorithms, allowing for faster evaluation of their performance benefits. This iterative development process facilitates fine-tuning and optimization of the GPU-accelerated code, leading to improved results and more efficient implementations.
5) **HLS for Custom Hardware Targets:** While GPUs offer significant parallelism and computational power, there may be scenarios where further acceleration or specific hardware targets (such as FPGAs or ASICs) are desired. By integrating CUDA with HLS tools, one gains the flexibility to explore alternative hardware targets. HLS tools automatically convert CUDA code into RTL descriptions that are suitable for hardware implementation on custom targets. This allows for exploring different hardware architectures, optimization techniques, and trade-offs to achieve even higher levels of performance, power efficiency, or specific design constraints.
6) **Faster Time-to-Market:** The combination of CUDA with HLS tools accelerates the development process and enables faster time-to-market for hardware-accelerated solutions. CUDA's productivity benefits, coupled with the option to explore alternative hardware targets using HLS, facilitate quicker algorithm development, optimization, and deployment. This advantage is particularly valuable in time-sensitive applications or industries where rapid innovation and efficient implementation are crucial.

3.2.2 OpenCL

OpenCL is not specifically designed for HLS. It is a parallel programming framework that enables developers to write code that can be executed on various devices, including CPUs, GPUs, and accelerators [20]. OpenCL provides a way to express parallel computations using its programming model, but it does not directly support HLS features like automated hardware generation.

3.2.2.1 OpenCL and HLS for Hardware Acceleration

OpenCL can be utilized in conjunction with HLS tools or frameworks to achieve hardware acceleration. Here is how it works as summarized in Refs. [21–23]:

1) **Algorithm Development:** Start by developing one's algorithm using the OpenCL programming language. OpenCL provides a platform-independent framework for writing parallel programs that can execute on various devices, including GPUs. Identify the computationally intensive parts of the algorithm that can benefit from parallel execution.
2) **OpenCL Implementation:** Implement the compute-intensive parts of the algorithm using OpenCL. OpenCL allows one to express parallel computations using kernels, which are functions that can be executed in parallel on multiple processing elements (PEs) available on the target hardware, such as GPUs. Within one's OpenCL code, one can define the work distribution across PEs, memory access patterns, and synchronization mechanisms.
3) **HLS Integration:** Once the compute-intensive parts in OpenCL is implemented, one can integrate OpenCL with HLS tools. HLS tools, such as Vivado HLS or Intel HLS Compiler, can analyze the OpenCL code and generate hardware descriptions in RTL format. RTL describes the behavior of individual hardware components and their interconnections.
4) **Hardware Synthesis:** With the HLS tools, one can perform hardware synthesis, which converts the RTL descriptions into a gate-level netlist. During synthesis, the HLS tools map the high-level OpenCL descriptions to specific hardware components, such as logic gates, multiplexers, and flip-flops. They also perform optimizations like resource sharing, pipelining, and scheduling to improve the performance, power consumption, and area of the design. The synthesis output is a detailed hardware description that can be further processed for implementation.
5) **Verification and Implementation:** Once the hardware design is synthesized, it needs to be verified to ensure correct functionality. This involves running test cases and comparing the results against the expected outputs to validate the correctness of the design. Additionally, performance metrics such as latency and throughput can be measured to evaluate the achieved acceleration. After successful verification, the design can be implemented on the target hardware platform, an FPGA or an ASIC. The implementation process involves mapping the gate-level netlist onto the physical resources of the target hardware, configuring the interconnections, and generating bitstreams or chip layouts for programming the FPGA or manufacturing the ASIC.

By integrating OpenCL with HLS tools, one can leverage the OpenCL's parallel computing capabilities for algorithm development and the hardware synthesis capabilities of HLS to transform the OpenCL code into RTL descriptions suitable for hardware implementation. This approach enables the acceleration of OpenCL-based algorithms on specific hardware targets like GPUs. The HLS tools provide optimizations, customization options, and mapping techniques to tailor the hardware design to the target hardware platform, maximizing performance and efficiency.

The advantages of using OpenCL with HLS for hardware acceleration are discussed in the next subsection.

3.2.2.2 Advantages of Using OpenCL with HLS Tools for Hardware Acceleration

Using OpenCL with HLS tools for hardware acceleration allows one to write platform-independent code and leverage the GPU's parallel processing capabilities. It also allows one to optimize and customize the hardware design as well as accelerate development and implementation processes for efficient, high-performance solutions. Here is a detailed explanation of the advantages.

1) **Hardware Portability:** OpenCL is a platform-independent framework that allows one to write parallel programs for heterogeneous systems, including GPUs. By utilizing OpenCL with HLS tools, one can develop hardware-accelerated designs that are portable across different platforms [20]. This flexibility enables one to target a wide range of hardware, including GPUs from various manufacturers, without having to rewrite the code for each specific platform.
2) **Productivity and Abstraction:** OpenCL provides a higher level of abstraction compared to lower-level HDLs used in HLS [23]. It allows one to express parallel computations using familiar C-like syntax, making writing, understanding, and maintaining parallel code easier. OpenCL abstracts away many low-level details of hardware design, such as memory access and synchronization, enabling developers to focus on the algorithm's logic rather than the intricacies of hardware implementation. This higher level of abstraction increases productivity and reduces the complexity associated with hardware acceleration.
3) **Leveraging GPU Parallelism:** GPUs are designed with a massive number of cores that can execute computations in parallel [24]. OpenCL allows one to harness the parallel processing power of GPUs by defining compute kernels that can be executed by multiple PEs simultaneously. By integrating OpenCL with HLS tools, one can optimize and customize the GPU-specific aspects of the design to exploit the full potential of parallelism offered by GPUs. This results in significant performance gains compared to running the same code on a CPU.

4) **HLS Optimizations:** HLS tools offer a range of optimizations to enhance the hardware design generated from OpenCL code. These optimizations include resource sharing, pipeline insertion, loop unrolling, and memory access optimizations [20]. The HLS tools analyze the OpenCL code and apply these optimizations automatically or with guidance from the developer. These optimizations aim to improve performance, reduce power consumption, and minimize the area required for the hardware implementation. By leveraging the optimization capabilities of HLS tools, one can achieve more efficient and optimized hardware designs.
5) **Customization for Specific Hardware Targets:** While GPUs provide substantial parallel processing capabilities, there may be scenarios where further acceleration or specific hardware targets, such as FPGAs or ASICs, are desired. By integrating OpenCL with HLS tools, one gains the flexibility to explore alternative hardware targets [21]. The HLS tools can analyze the OpenCL code and generate hardware descriptions suitable for the target platform, enabling one to customize the hardware design to meet specific requirements. This customization allows one to achieve higher levels of performance, power efficiency, or target-specific design constraints.
6) **Faster Time-to-Market:** The combination of OpenCL with HLS tools enables faster time-to-market for hardware-accelerated solutions [21]. OpenCL's platform independence and hardware portability allow for quicker development and testing across a range of platforms. The productivity gains provided by OpenCL's higher-level abstraction and the optimization capabilities of HLS tools expedite the algorithm development, optimization, and implementation processes. This advantage is particularly valuable in time-sensitive applications or industries where rapid innovation and efficient implementation are crucial.

In summary, OpenCL itself is not an HLS language. It is a parallel programming framework that can be used in conjunction with HLS tools or frameworks to program and execute hardware designs generated from high-level code. The combination of HLS and OpenCL allows developers to take advantage of automated hardware generation and the flexibility of programming heterogeneous platforms.

3.2.3 Maxeler MaxJ

Maxeler MaxJ is an HLS language developed by Maxeler Technologies [25]. It enables programmers to write high-level code that is automatically transformed into HDL code for hardware synthesis. MaxJ follows a dataflow programming model, allowing computations to be expressed as interconnected dataflow graphs [26]. It provides features such as a rich type system, memory hierarchy control, streaming, and pipelining constructs to optimize hardware utilization and performance. MaxJ also offers performance analysis tools to assist developers in identifying bottlenecks and optimizing hardware designs.

Maxeler MaxJ serves as an effective HLS language by offering a dataflow programming model, precise type control, and memory management capabilities. Programmers can express computations as dataflow graphs, enabling optimized resource utilization during hardware synthesis. The type system allows for fine-grained control over data representation, striking a balance between hardware resources and numerical precision. Additionally, MaxJ provides constructs for managing on-chip memory efficiently, minimizing data transfers between external memory and the hardware circuit.

Streaming and pipelining support in MaxJ facilitates efficient data movement and parallel execution, thus reducing latency and maximizing the throughput. These features are complemented by performance analysis tools that help developers understand the performance characteristics of their hardware designs and guide optimization efforts. In summary, Maxeler MaxJ, as an HLS language, enables programmers to express high-level computations and automatically generate efficient hardware designs with improved performance and resource utilization.

It is important to note that Maxeler MaxJ is a specialized HLS language tailored specifically for dataflow computing. It may have a learning curve and require familiarity with the dataflow programming model. The performance benefits and productivity gains of MaxJ depend on the nature of the application and how well it can be expressed and parallelized as dataflow computations. However, for applications that can be efficiently modeled and parallelized in a dataflow manner, Maxeler MaxJ can provide significant performance advantages and productivity improvements in hardware design. We will discuss how Maxeler MaxJ can be used with HLS for hardware acceleration.

3.2.3.1 Using Maxeler MaxJ with HLS for Hardware Acceleration

By utilizing Maxeler MaxJ with HLS, one can describe the algorithm in a high-level, dataflow-based programming model, allowing for efficient hardware acceleration. The MaxJ compiler (MaxCompiler) performs HLS, generating an optimized hardware design from the MaxJ code [27]. The resulting hardware implementation can be deployed on a Maxeler dataflow engine (DFE) or an FPGA. The dataflow-based execution model, combined with parallelism and pipelining optimizations, enables high-performance computing on specific hardware targets like GPUs. Here is a detailed description of the process based on Refs. [28, 29].

1) **Algorithm Description in MaxJ:** To begin, one describe one's algorithm using the MaxJ language. MaxJ provides constructs for expressing parallelism, data dependencies, and pipelining. One defines a dataflow graph that represents the flow of data between different operations or compute kernels. The

dataflow graph allows the compiler to understand the dependencies between operations and optimize the implementation for efficient parallel execution.

2) **HLS Compilation with MaxCompiler:** Once one has written the MaxJ code, one uses the MaxCompiler, which is the HLS toolset provided by Maxeler Technologies. The MaxCompiler takes one's MaxJ code as input and performs analysis and optimization. It generates an optimized hardware design based on the specified dataflow graph and other directives provided in the MaxJ code.
3) **Hardware Generation:** The MaxCompiler performs HLS to convert the MaxJ code into a hardware description. It maps the operations and data dependencies defined in the MaxJ code onto specific hardware components, such as PEs, memory banks, and interconnects. The generated hardware description is typically in an HDL format, such as VHDL or Verilog.
4) **Compilation and Deployment:** The hardware description generated by the MaxCompiler can be compiled and synthesized using standard HDL tools. This process transforms the hardware description into a bitstream or configuration file that can be loaded onto the target hardware. The target hardware can be an FPGA or a custom Maxeler DFE, a specialized hardware platform provided by Maxeler Technologies.
5) **Hardware Acceleration:** Once the hardware design is deployed on the target hardware, the Maxeler DFE or FPGA executes the algorithm in parallel. The hardware implementation directly executes the dataflow graph specified in the MaxJ code. It takes advantage of the inherent parallelism and pipelining defined in the MaxJ code to achieve significant speedup over traditional software execution. The hardware implementation can exploit parallel processing capabilities, including those found in GPUs, by offloading specific computation-intensive tasks to the GPU while still maintaining the dataflow-based acceleration provided by the Maxeler DFE.
6) **Data Transfer and Collaboration:** In the case of GPU acceleration, Maxeler's DFE can interface with the GPU to exchange data between the GPU and the DFE. This allows for collaboration between the two hardware platforms, where the DFE offloads computation-intensive tasks to the GPU for parallel processing while benefiting from the dataflow-based acceleration provided by the DFE. This combination of the DFE and GPU can further enhance the overall performance and accelerate the algorithm's execution.

3.2.3.2 Advantages of Using Maxeler MaxJ with HLS for Hardware Acceleration

Using Maxeler MaxJ with HLS tools provides advantages such as a dataflow programming model, high-level abstraction, HLS optimization capabilities, platform adaptability, performance and energy efficiency, rapid prototyping and iteration, scalability, and reusability. These advantages streamline the development process, optimize hardware designs, and enable efficient hardware acceleration for a wide range of applications. The following points summarize these advantages [29, 30].

1) **Dataflow Programming Model:** Maxeler MaxJ is designed around a dataflow programming model, which allows one to express algorithms based on data dependencies rather than control flow. This model simplifies the programming process and provides a natural representation for many parallel algorithms. By describing one's algorithm in MaxJ, one can take advantage of dataflow optimizations and parallelism to achieve efficient hardware acceleration.
2) **High-Level Abstraction:** Maxeler MaxJ provides a high-level abstraction for hardware design, allowing one to focus on the algorithm's logic rather than low-level implementation details. The Java-based syntax makes it accessible to programmers familiar with Java or similar languages, reducing the learning curve. By abstracting away hardware complexities, MaxJ enables faster development and easier maintenance of hardware-accelerated designs.
3) **HLS Optimization Capabilities:** Maxeler MaxJ integrates with HLS tools such as the MaxCompiler, which perform analysis and optimization on the MaxJ code. The HLS tools apply various optimizations, including parallelization, pipelining, loop unrolling, and memory optimization, to improve the performance, latency, throughput, and resource utilization of the resulting hardware design. These optimizations are performed automatically or with guidance from the developer, ensuring that the hardware design is highly optimized for the target platform.
4) **Platform Adaptability:** Maxeler MaxJ is designed to work with different hardware platforms, including Maxeler DFEs and FPGAs. The MaxCompiler generates hardware descriptions that can be compiled and synthesized for these platforms. This platform adaptability allows one to choose the hardware target that best suits one's performance, power, and cost requirements. It also enables one to explore different hardware options without major code modifications, reducing development time and effort.
5) **Performance and Energy Efficiency:** By leveraging dataflow-based execution and the optimizations provided by HLS tools, Maxeler MaxJ enables hardware designs that deliver high-performance acceleration. The dataflow model enables parallelism and pipelining, maximizing resource utilization and throughput. The HLS optimizations further enhance performance, reducing latency, and maximizing the utilization of hardware resources. Additionally, the efficient use of parallelism and optimized resource allocation in the hardware design contribute to improved energy efficiency.
6) **Rapid Prototyping and Iteration:** Maxeler MaxJ's high-level abstraction and HLS integration facilitate rapid prototyping and iteration. The ability to express algorithms at a higher level of abstraction reduces development time and allows for quick evaluation and refinement of hardware-accelerated designs. The HLS

tools streamline the design process, enabling faster exploration of different optimization strategies, architecture configurations, and platform targets.

7) **Scalability and Reusability:** Maxeler MaxJ supports scalable hardware designs that can be easily adapted to handle larger datasets or more complex algorithms. The dataflow model allows easy scalability by adding or modifying dataflow nodes to accommodate changing requirements. Additionally, MaxJ promotes modular design and code reuse, making it easier to reuse and adapt components across different projects or use cases. This scalability and reusability contribute to efficient development, code maintenance, and future-proofing of hardware-accelerated solutions.

3.3 Limitations and Challenges of HLS

Despite the great benefits and advantages of HLS, it has some limitations and challenges. Some of the limitations could be overcome through some identified workarounds. However, developers must carefully decide whether to use or proceed with the HLS in scenarios with obvious or undesirable limitations. Here are some of the limitations and challenges of HLS:

a) **Hardware System Dependence:** HLS tools do not generate complete hardware-independent designs. Usually, developers have to tailor their designs and direct the HLS tool toward their targeted hardware by giving it clear and appropriate instructions relevant to the targeted hardware. Therefore, developers still need to have some basic understanding of their target hardware.
b) **Lack of Complete Hardware Feature Support:** Some hardware features, such as memory interfaces, distributed memory model, custom IPs, and timing constraints in FPGAs, are not directly supported in HLS. Developers may need to use workarounds, such as using external wrappers or libraries, for some features when feasible. In some cases, a few lines of HDL are embedded in the HLS.
c) **Unsupported Software Features:** Contrary to the hardware feature limitation, in this case, the target hardware does not directly support some software features such as dynamic memory allocation and pointer. Therefore, it becomes difficult to implement such features efficiently using HLS.
d) **Overheads in Comparison to HDL:** Although HLS speeds up time-to-market, increases developers' productivity and efficiency, and achieves other benefits compared to HDL, the HLS brings some overheads compared to HDL. HLS-generated codes may have higher resource utilization and consume more power than well-written HDL codes. The performance of HLS-generated codes is usually lower than that of a well-written and optimized HDL, even though the fast development in HLS tools may soon fill this gap.

e) **Expertise Requirement:** Although HLS development is much easier than HDL hardware development, HLS still requires some skills and expertise on top of pure software development. For example, an HLS developer is required to have a basic understanding of target hardware architecture, application-specific hardware design, and some optimization skills. Therefore, HLS does not eliminate the knowledge of hardware design for an efficient design.

In summary, HLS not only offers several benefits over traditional HDL but also has some limitations compared with HDL. Therefore, it is paramount for the developer to know the best time to use HLS or HDL. Developers may opt for the HDL for systems having complex resource constraints and extreme performance requirements. Tasks that use dynamic memory allocations and pointers may not be suitable for hardware acceleration with HLS. In contrast, HLS is highly suitable for implementing complex mathematical procedures such as FFT, NTT, security primitives, and other digital signal processing primitives. HLS is also suitable for improving the developer's productivity, reducing time-to-market, design-space exploration, algorithm optimization, and ASIC developments.

3.4 Using HLS in 5G Edge Computing

HLS has many applications. One example application is IC design, in which HLS converts high-level abstractions to Verilog. With the recent trend of data processing acceleration, heterogeneous computing is used. Hardware assists software in the acceleration of data processing. For example, specialized hardware such as GPU can be used to assist the CPU for applications such as ChatGPT. It is important to identify the bottleneck processing areas where acceleration is needed before assigning the specialized hardware, such as GPU, ASICs, or FPGAs to them. Heterogeneous computing can be used in 5G edge computing to accelerate both the user and control plane functions.

HLS is beneficial for 5G edge computing in several ways. First, HLS enables the performance optimization by allowing algorithm designers to express their algorithms in high-level languages and automatically generate efficient hardware designs. This is particularly important for tasks such as signal processing, data compression, encryption/decryption, and real-time analytics in 5G edge computing [31–36].

Second, HLS reduces time-to-market by accelerating the development cycle [31, 36]. Designers can quickly convert high-level code into hardware designs, minimizing the need for manual hardware design and verification. This enables faster deployment of new services and applications in the dynamic 5G edge computing environment.

Additionally, HLS offers flexibility and adaptability, allowing designers to easily modify and customize their designs to meet changing requirements and evolving 5G standards. It also helps in achieving power and energy efficiency by providing optimizations for power consumption, enabling designers to explore low-power design techniques and power management strategies [32, 33, 37].

Furthermore, HLS supports scalability and customization, enabling designers to explore scalable hardware architectures suitable for edge computing and tailor the hardware designs to specific applications and constraints at the network edge [31, 35].

By leveraging HLS in 5G edge computing, designers can achieve performance optimizations, power efficiency, real-time processing capabilities, flexibility, and seamless integration with software components. These benefits contribute to the efficient deployment of edge computing platforms, enabling low-latency, high-throughput processing, and real-time decision-making in 5G networks at the network edge. In the following sections, we will discuss how HLS can support 5G edge computing via the user plane and control plane in the following sections.

3.4.1 User (Data) Plane Acceleration

HLS can significantly support 5G edge computing through user plane acceleration. The user plane handles data transmission and processing in 5G networks. HLS can enhance user plane acceleration in the following ways:

1) **Hardware Acceleration:** HLS allows designers to create custom hardware accelerators specifically optimized for the computationally intensive tasks in the user plane [32]. By describing these tasks at a high level, HLS tools generate hardware designs that exploit parallelism, optimize memory accesses, and maximize resource utilization. This hardware acceleration offloads processing from the general-purpose processors, enabling faster and more efficient execution of tasks such as packet forwarding, encryption/decryption, quality-of-service (QoS) management, and traffic shaping. By utilizing dedicated hardware accelerators, the user plane can handle higher data rates, reduce latency, and achieve higher throughput.
2) **Protocol Offloading:** 5G networks involve complex protocol stacks including the NR protocol stack, which require efficient processing and analysis. HLS enables designers to express the protocol processing tasks at a high level, allowing the generation of hardware designs that specialize in these protocols [37]. By offloading the protocol processing to dedicated hardware accelerators, the user plane can efficiently parse, classify, and analyze network packets. This offloading frees up CPU resources and reduces the processing burden on the general-purpose processors, resulting in improved system performance and reduced latency.

3) **Real-time Processing:** HLS plays a crucial role in supporting real-time processing requirements in the user plane of 5G edge computing [38]. HLS tools allow designers to express the timing requirements and constraints of real-time applications at a high level. The generated hardware designs are optimized to meet these timing constraints, ensuring the timely processing of critical data. Real-time processing capabilities enable the user plane to handle time-sensitive tasks such as low-latency video streaming, real-time analytics, and responsive control systems. By leveraging HLS, the user plane can meet stringent latency requirements and deliver high-quality, real-time services.

In addition, HLS can perform the following two tasks for both the user and control planes.

4) **Custom Data Path Implementation:** HLS enables designers to implement custom data paths tailored to the specific requirements of the user and data plane tasks. By expressing the data path at a high level, HLS tools generate hardware designs that efficiently process and route data through optimized pipelines [39]. This customization allows designers to optimize the data flow, minimize data movement, and maximize parallelism. For example, in packet processing, HLS can generate hardware designs efficiently that perform header parsing, payload inspection, and flow classification. Custom data path implementations enhance performance, reduce latency, and increase the throughput of the user plane and control plane.
5) **Integration with Software:** HLS facilitates seamless integration between hardware accelerators and the software stack in 5G edge computing [40]. HLS tools provide frameworks, interfaces, and APIs that enable smooth communication and data exchange between the generated hardware designs and the software running on the edge devices. This integration allows efficient coordination between the hardware accelerators and the software components in the user and data planes. It enables collaborative processing, where the software can configure and control the hardware accelerators, exchange data with them, and coordinate their operations. Integrating the hardware and software components ensures efficient utilization of resources, synchronization, and interoperability in the user and data planes.

3.4.2 Control Plane Acceleration

The control plane is responsible for handling and processing network data traffic in 5G networks. HLS can enhance control plane acceleration in the following ways:

1) **Hardware Acceleration:** HLS enables the design and implementation of custom hardware accelerators highly optimized for control plane tasks [37].

These hardware accelerators are designed to efficiently handle specific operations, such as user authentication, access management, and policy control. These operations require massive encryption, decryption, and digital signature executions, which will significantly benefit from hardware acceleration. By describing these tasks at a high level, HLS tools automatically generate hardware designs that exploit parallelism, optimize memory access patterns, and maximize resource utilization. Hardware acceleration offloads these computationally intensive operations from the general-purpose processors, enabling faster and more efficient access control in the control plane. This acceleration improves the overall throughput and reduces latency in 5G networks.

2) **Traffic Management:** HLS supports the acceleration of traffic management functions in the control plane, which is crucial for ensuring efficient and reliable data transmission in 5G networks [40]. Traffic management tasks include traffic shaping, congestion control, QoS management, and flow control. By describing these tasks at a high level, HLS enables the generation of hardware designs that efficiently handle traffic management operations. These designs can implement algorithms and mechanisms for prioritizing traffic, managing bandwidth allocation, detecting and mitigating congestion, and enforcing QoS policies. Hardware acceleration of traffic management tasks improves network performance, reduces packet loss, and ensures optimal resource utilization in the control plane.

By leveraging HLS for user and control planes acceleration in 5G edge computing, designers can achieve hardware acceleration, efficient packet processing, traffic management, custom data path implementations, and seamless integration with software components. These capabilities collectively enhance the performance, efficiency, and responsiveness of the user and control planes, enabling efficient handling of network data traffic, reducing latency, and improving user experiences in edge computing environments. HLS empowers designers to optimize the user and control planes for demanding requirements.

3.4.3 Advantages of Using HLS for User Plane and Control Plane Acceleration

Here is a detailed description of the advantages of using HLS for the user and control planes acceleration.

1) **Performance Enhancement:** HLS enables the acceleration of user plane or control plane processing tasks in 5G edge computing systems, leading to significant performance improvements. By utilizing HLS, computationally intensive operations, such as packet processing, encryption/decryption, traffic classification, or flow analysis, can be offloaded to a specialized hardware [33, 34]. HLS

tools automatically generate optimized hardware designs that leverage parallelism, pipelining, and other optimizations to achieve higher throughput and lower latency. This acceleration of data plane processing tasks allows for faster and more efficient handling of large volumes of data, improving the overall performance of 5G edge computing systems.

2) **Lower Latency:** HLS allows for the optimization of hardware designs to reduce latency in the user or control plane of 5G edge computing [31, 41]. By leveraging HLS, designers can specify and optimize the hardware architecture to minimize data path delays, maximize parallelism, and ensure efficient data flow. This reduction in latency is crucial for applications requiring real-time responsiveness, such as voice and video communication, gaming, or real-time analytics. By accelerating data plane processing tasks through HLS, the overall latency of the system can be minimized, enabling near-instantaneous data processing and enabling low-latency applications to operate effectively.
3) **Power Efficiency:** HLS can contribute to lower power consumption and improved energy efficiency in 5G edge computing systems [32, 33, 37]. By optimizing the hardware design through HLS, power-efficient implementations can be achieved. HLS tools apply techniques such as resource sharing, clock gating, or power-aware synthesis to minimize power consumption in the resulting hardware. By offloading compute-intensive tasks to dedicated hardware through HLS, power requirements in the data plane can be reduced, leading to more efficient energy utilization and longer battery life in edge devices. This power efficiency is crucial for energy-constrained edge environments where minimizing power consumption is a priority.
4) **Design Productivity:** HLS significantly enhances design productivity for user and control planes acceleration in 5G edge computing. By raising the level of abstraction and automating the translation of high-level descriptions into hardware designs, HLS simplifies the design process [37, 42]. Designers can express their designs using high-level languages, thus leveraging their existing software development skills. This integration of software and hardware design reduces the design effort, eliminates the need for manual hardware design from scratch, and allows for faster exploration of design alternatives. HLS enables rapid prototyping, iterative design, and faster time-to-market for 5G edge computing solutions.
5) **Flexibility and Adaptability:** HLS provides designers with flexibility and adaptability in the design of user and control planes acceleration for 5G edge computing [37]. Designers can easily modify and experiment with the high-level descriptions, enabling exploration of different architectural configurations, algorithms, or design parameters. This flexibility allows for efficient design space exploration, optimization for specific performance metrics, and customization for different use cases and network conditions. HLS enables

designers to adapt their designs to evolving 5G requirements and rapidly prototype and iterate on data plane acceleration solutions, ensuring optimal performance and efficiency.

6) **Reusability:** HLS promotes the reusability of designs for user and control planes acceleration in 5G edge computing. HLS-generated hardware designs are often portable and can be retargeted to different hardware platforms [35]. This reusability allows designers to leverage existing IP and design modules, reducing development effort and time-to-market. By reusing validated and optimized components, designers can focus on specific functionalities or optimizations required for data plane acceleration, speeding up the development process and ensuring the reliability and maturity of the design. This reusability also enables efficient sharing of hardware accelerators across different applications and platforms, promoting modularity and scalability in 5G edge computing.

References

1 Schaefer, B.C. (2023). High-level synthesis made easy, highX Technologies. www.hlsbook.com (accessed on 1 June 2023).

2 Martin, G. and Smith, G. (2009). High-level synthesis: past, present, and future. *IEEE Design & Test of Computers* 26 (4): 18–25.

3 Krishnapriya, P.N. and Sundari, B.B.T. (2018). High level synthesis for retiming stochastic VLSI signal processing architectures. *Procedia Computer Science* 143: 10–19.

4 Namballa, R., Ranganathan, N., and Ejnioui, A. (2004). Control and data flow graph extraction for high-level synthesis. *IEEE Computer Society Annual Symposium on VLSI,* 4 October 2004, Lafayette LA USA, pp. 187–192. IEEE.

5 Gupta, R. and Brewer, F. (2008). High-level synthesis: a retrospective. In: Coussy, P. and Morawiec, A. (Eds). *High-Level Synthesis: From Algorithm to Digital Circuit*, 13–28. Springer.

6 McFarland, M.C., Parker, A.C., and Camposano, R. (1990). The high-level synthesis of digital systems. *Proceedings of the IEEE* 78 (2): 301–318.

7 Cong, J., Liu, B., Neuendorffer, S. et al. (2011). High-level synthesis for FPGAs: from prototyping to deployment. *IEEE Transactions on Computer-Aided Design of Integrated Circuits and Systems* 30 (4): 473–491.

8 Coussy, P., Gajski, D.D., Meredith, M., and Takach, A. (2009). An introduction to high-level synthesis. *IEEE Design & Test of Computers* 26 (4): 8–17.

9 Numan, M.W., Phillips, B.J., Puddy, G.S., and Falkner, K. (2020). Towards automatic high-level code deployment on reconfigurable platforms: a survey of high-level synthesis tools and toolchains. *IEEE Access* 8: 174692–174722.

10 Meeus, W., Van Beeck, K., Goedemé, T. et al. (2012). An overview of today's high-level synthesis tools. *Design Automation for Embedded Systems* 16: 31–51.

11 Siemens. Catapult high level synthesis overview. (https://eda.sw.siemens.com/en-US/ic/catapult-high-level-synthesis), 2024. Accessed on 1 September 2024.

12 Cadence. Stratus high level synthesis overview. (https://www.cadence.com/en_US/home/tools/digital-design-and-signoff/synthesis/stratus-high-level-synthesis.html), 2024. Accessed on 1 September 2024.

13 Canis, A., Choi, J., Aldham, M. et al. (2013). LegUp: an open-source high-level synthesis tool for FPGA-based processor/accelerator systems. *ACM Transactions on Embedded Computing Systems (TECS)* 13 (2): 1–27.

14 Nvidia. CUDA Zone. (https://developer.nvidia.com/cuda-zone), 2024. Accessed on 1 September 2024.

15 Nickolls, J., Buck, I., Garland, M., and Skadron, K. (2008). Scalable parallel programming with CUDA: is CUDA the parallel programming model that application developers have been waiting for? *Queue* 6 (2): 40–53.

16 Nickolls, J. and Dally, W.J. (2010). The GPU computing era. *IEEE Micro* 30 (2): 56–69.

17 Mu, Q., Cui, L., and Song, Y. (2015). The implementation and optimization of Bitonic sort algorithm based on CUDA. *arXiv preprint*. arXiv:1506.01446.

18 Hoydis, J., Cammerer, S., Aoudia, F.A. et al. (2022). Sionna: an open-source library for next-generation physical layer research. *arXiv preprint*. arXiv:2203.11854.

19 Zarif, M.I.I., Shahria, M.T., Sunny, M.S.H., and Rahaman, M.M. (2022). A vision-based object detection and localization system in 3D environment for assistive robots' manipulation. *Proceedings of the 9th International Conference of Control Systems, and Robotics (CDSR'22)*, 2–4 June 2022, Niagara Falls Canada, pp. 2–4.

20 Khronos OpenCL Working Group (2021). The OpenCL specification.

21 Muslim, F.B., Ma, L., Roozmeh, M., and Lavagno, L. (2017). Efficient FPGA implementation of OpenCL high-performance computing applications via high-level synthesis. *IEEE Access* 5: 2747–2762.

22 Coole, J. and Stitt, G. (2013). Fast, flexible high-level synthesis from OpenCL using reconfiguration contexts. *IEEE Micro* 34 (1): 42–53.

23 Venkatakrishnan, R., Misra, A., and Kindratenko, V. (2020). High-level synthesis-based approach for accelerating scientific codes on FPGAs. *Computing in Science & Engineering* 22 (4): 104–109.

24 Daoud, L., Zydek, D., and Selvaraj, H. (2014). A survey of high level synthesis languages, tools, and compilers for reconfigurable high performance computing. *Advances in Systems Science: Proceedings of the International Conference on Systems Science 2013 (ICSS 2013)*, 10–12 September 2013, Wroclaw Poland, pp. 483–492. Springer International Publishing.

25 Maxeler Technologies Home Page. https://maxeler.com/. Accessed on 1 September 2024.

26 Kritikakis, C. and Koch, D. (2020). Enabling dynamic system integration on Maxeler HLS platforms. *Journal of Signal Processing Systems* 92: 887–905.

27 Maxeler Technologies Maxcompiler User Manual. https://www.maxeler.com/products/software/maxcompiler. Accessed on 1 September 2024.

28 Todman, T. and Luk, W. (2017). In-circuit assertions and exceptions for reconfigurable hardware design. *Provably Correct Systems* 265–281.

29 Milutinović, V. (2014). Dataflow supercomputing: concepts of importance. *ETF Journal of Electrical Engineering* 20 (1): 146–156.

30 Korolija, N., Djukic, T., Milutinovic, V., and Filipovic, N. (2013). Accelerating lattice-Boltzman method using Maxeler dataflow approach. *The IPSI BgD Transactions on Internet Research* 34–42. http://ipsitransactions.org/journals/papers/tir/2013july/tir2013july.pdf#page=36.

31 Gerogiannis, G., Birbas, M., Leftheriotis, A. et al. (2022). Deep reinforcement learning acceleration for real-time edge computing mixed integer programming problems. *IEEE Access* 10: 18526–18543.

32 Shah, N.A., Lazarescu, M.T., Quasso, R. et al. (2022). FPGA acceleration of 3GPP channel model emulator for 5G new radio. *IEEE Access* 10: 119386–119401.

33 Sampath, K.J., Kumar, N.K., Yeswanth, K. et al. (2021). An efficient channel coding architecture for 5G wireless using high-level synthesis. *2021 5th International Conference on Trends in Electronics and Informatics (ICOEI)*, 3–5 June 2021, Tirunelveli India, pp. 674–680. IEEE.

34 Sikka, P., Asati, A.R., and Shekhar, C. (2020). Speed optimal FPGA implementation of the encryption algorithms for telecom applications. *Microprocessors and Microsystems* 79: 103324.

35 Mhaske, S., Kee, H., Ly, T. et al. (2017, 2017). FPGA-based channel coding architectures for 5G wireless using high-level synthesis. *International Journal of Reconfigurable Computing* 3689308:1–3689308:23.

36 Choi, G., Park, K.B., and Chung, K.S. (2018). Optimization of FPGA-based LDPC decoder using high-level synthesis. *Proceedings of the 4th International Conference on Communication and Information Processing*, 2 November 2018, Qingdao Shandong China, pp. 256–259.

37 Civerchia, F., Pelcat, M., Maggiani, L. et al. (2020). Is OpenCL driven reconfigurable hardware suitable for virtualising 5G infrastructure? *IEEE Transactions on Network and Service Management* 17: 849–863.

38 Benacer, I., Boyer, F.R., and Savaria, Y. (2019). A high-speed, scalable, and programmable traffic manager architecture for flow-based networking. *IEEE Access* 7: 2231–2243.

39 Ricart-Sanchez, R., Malagon, P., Salva-Garcia, P. et al. (2018). Towards an FPGA-accelerated programmable data path for edge-to-core communications in 5G networks. *Journal of Network and Computer Applications* 124: 80–93.

40 Bose, A., Kirtikar, S., Chirumamilla, S. et al. (2022). AccelUPF: accelerating the 5G user plane using programmable hardware. *Proceedings of the Symposium on SDN Research*, 19–20 October 2022, Virtual Conference, pp. 1–15.

41 Silva, J.S., Boyer, F.R., and Langlois, J.M. (2017). P4-Compatible high-level synthesis of low latency 100 Gb/s streaming packet parsers in FPGAs. *Proceedings of the 2018 ACM/SIGDA International Symposium on Field-Programmable Gate Arrays*, 15 February 2018, Monterey California USA, pp. 147–152.

42 Gao, X., Raghunathan, D., Fang, R. et al. (2022). High-level synthesis for packet-processing pipelines. *ArXiv*. abs/2211.06475.

4

Coding Design

4.1 Overview

Data integrity and security are essential aspects of 5G networks' security and reliability.

Data integrity refers to the accuracy and completeness of data transmitted over the network. In other words, it ensures that data is not modified or corrupted during transmission. 5G networks use various techniques such as error correction codes (ECCs) and cryptographic algorithms to ensure data integrity. ECCs allow the receiver to detect and correct errors in the transmitted data, while cryptographic algorithms ensure that data is not tampered with during transmission.

Data security in wireless networks such as 5G refers to the measures and techniques used to protect the confidentiality, integrity, and availability of data transmitted over the network. Data security is critical in a 5G network because of the large volume of data transmitted and the high speed at which it is transmitted.

There are several key components of data security in 5G networks, including:

a) **Encryption:** Encryption involves using algorithms to convert data into a format that can only be read by authorized users with the correct decryption keys. In a 5G network, encryption are used to protect data as it is transmitted between devices and over the network.
b) **Authentication**: Authentication involves verifying the identity of a user or device before allowing access to the network. In a 5G network, authentication is used to prevent unauthorized access to the network and to ensure that data is only accessed by authorized users.
c) **Access Control:** Access control involves using policies and controls to restrict access to network resources based on user identity, device type, and other factors. In a 5G network, access control is used to prevent unauthorized access to network resources and ensure that data is only accessed by authorized users.

Edge Computing Acceleration: From 5G to 6G and Beyond, First Edition.
Patrick Hung, Hongwei Kan, and Greg Knopf.

d) **Intrusion Detection and Prevention:** Intrusion detection and prevention involves using software and hardware tools to detect and prevent unauthorized access to the network. In a 5G network, intrusion detection and prevention is used to detect and prevent attacks such as hacking, malware, and denial-of-service (DoS) attacks.
e) **Data Loss Prevention:** Data loss prevention involves using policies and controls to prevent the loss or theft of sensitive data. In a 5G network, data loss prevention is used to prevent the loss or theft of sensitive data as it is transmitted over the network.

Overall, data integrity and security in a 5G network are essential to ensure that data is protected as it is transmitted over the network. By using a range of techniques such as ECCs, encryption, authentication, access control, intrusion detection and prevention, and data loss prevention, organizations can ensure that their data is protected from unauthorized access, theft, and other security threats. In this chapter, we will discuss some key coding schemes and encryption algorithms.

4.2 Error Correction Codes (ECCs)

Compared to 4G, 5G communication systems have very significant performance improvements. The established standard of 5G New Radio (NR) demands high throughput, e.g., the peak data download rate can reach 20 Gbps, low decoding latency, e.g., the end-to-end connection data delay is less than 1 ms, and high reliability, e.g., the reliability is 99.999%. Hence, 3GPP has adopted two types of state-of-the-art ECC schemes (low-density parity-check [LDPC] codes and Polar codes) for 5G NR to satisfy these vital requirements.

Due to noise, fading, or interference, data transmitted over unreliable channels may contain errors. An ECC is one of the key technologies to protect the accuracy of data in modern communication and storage systems. The central idea of an ECC is to add the redundancy information with the raw message at the transmitter, and then the receiver can use the redundancy to detect and correct a limited number of errors that may occur during transmission. The noisy-channel coding theorem provided by Shannon [1] gives the theoretical minimum redundancy messages, which the most effective ECC must use to ensure the error probability can turn to zero. Reliable communication over a channel requires the information transfer rate R_b to be less than the capacity C of this channel. For example, for the continuous-time Additive White Gaussian Noise (AWGN) channel bandlimited to W Hertz, reliable communication is possible provided that

$$R_b < C = W \quad \log_2 \left(1 + \frac{S}{N}\right) \text{ bits/s}$$

where S/N is the signal-to-noise ratio (SNR).

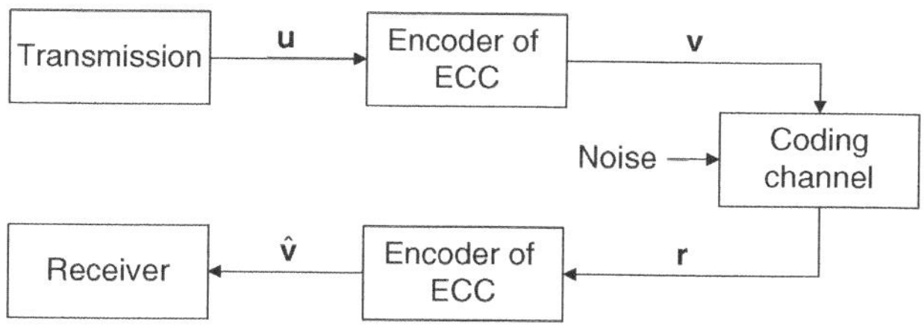

Figure 4.1 Simplified model of an ECC system.

An ECC in the transmitter and receiver is, respectively, referred to as the encoder and decoder. A simplified model of an ECC system is given as follows, where **u** represents the raw message sequence, **v** represents the encoded sequence, **r** is the output sequence of the coding channel, which is called the received sequence, and $\hat{\mathbf{v}}$ represents the decoded sequence (Figure 4.1).

ECCs are widely applied in situations like one-way communication systems, where it is hard to request re-transmissions or without a reverse channel. Different ECC schemes have different redundant ways, thus the maximum number of errors or erasures (i.e., missing messages) that can be corrected are also different. In general, the more redundancy messages are transmitted (i.e., the lower the information transfer rate), the more errors or erasures can be corrected. Thus, ECCs are at the cost of available bandwidth rather than the effective SNR to protect the reliability of data transmitted over a noisy channel. The performance of ECCs can be characterized by the SNR or the block error rate (BLER) of the decoder. Good ECCs can ensure the receiver can identify the transmitted original messages from the received message without re-transmission but with reasonable overhead of channel bandwidth [2]. Although the noisy-channel coding theorem did not give the constructions of the most effective ECCs, in the last 20 years, some channel capacity-approaching or -achieving codes have been found, such as Turbo codes, LDPC codes, and Polar codes.

In general, ECCs can be categorized into two types according to their structures: block codes and convolution codes. The input message for the encoder of a block code is a message block **u** of k discrete bits, and the output block **v** of the encoder is obtained by independently encoding the corresponding message block **u**. The output block **v** contains n discrete bits or symbols and is called as one codeword of the block code. Since the encoder of a block code is memoryless, it can be implemented by a combinational logic circuit. An (n,k) binary block code has 2^k different possible codewords of length n, associated with 2^k possible message blocks of length k. The code rate of an (n,k) block code is the ratio $R = k/n$. Some of the classical block codes are:

- Hamming codes
- Golay codes

- Reed–Muller (RM) codes
- Product codes
- Bose–Chaudhuri–Hocquenghem (BCH) codes
- Reed–Solomon (RS) codes

Unlike block codes, the encoder of a convolutional code contains memory. Thus, each codeword sequence **v** of length k is determined not only by the current input message block **u** but also by the m previous message bits. Hence, the encoder of a convolutional code must be implemented by a sequential logic circuit, and m is called memory order. The code rate of an (n,k,m) convolutional code is $R = k/n$.

Classical block codes, such as BCH and RS codes, have their error-detecting and -correcting capabilities specified by an important parameter – minimum Hamming distance d. The minimum Hamming distance is the smallest Hamming distance of two code words when the message code words are paired with one another. To detect up to e error bits, the minimum Hamming distance d of a block code satisfies $d \geq e + 1$. Likewise, to correct up to e error bits, the minimum Hamming distance d must be $d \geq 2e + 1$. The minimum Hamming distance of classical block codes can be increased by increasing the number of k and n to detect and correct more errors. Similarly, the error-detecting and -correcting capabilities of convolutional codes can be determined by an important parameter – free distance. By increasing the number of m rather than the number of k and n, the free distance of convolutional codes can be increased to achieve lower BLER.

4.2.1 Turbo, Low-Density Parity-Check, and Polar Codes

The classical ECC codes have many structures, such as algebraic and topological structures. These structures can guarantee good minimum distance properties. However, the structured ECC codes usually cannot approach the performance promised by Shannon, even if for sufficiently long block lengths. This is because the channel capacity-approaching codes require random properties to obtain the best distance properties. However, due to the lack of structures, it is too difficult to decode random codes, and the complexity of the maximum likelihood decoding (MLD) for random codes of sufficiently long block lengths is too high to actually actually implement. Hence, researchers of the last century have struggled to design capacity-approaching codes with implementable decoding algorithms; however, a 2–3 dB gap always exists between their gain and the Shannon limit.

Until the 1990s, channel capacity-approaching codes, such as Turbo, LDPC, and Polar codes with modest decoding complexity, were invented or rediscovered successively [3, 4]. Unlike the classical ECC codes, these modern codes are random-like, mainly evaluated in terms of BLERs rather than minimum distances. This chapter will briefly introduce three famous channel capacity-approaching or -achieving codes.

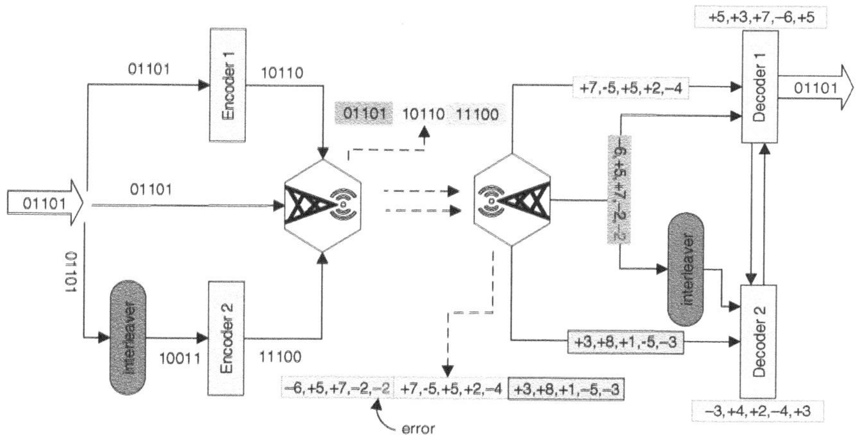

Figure 4.2 Encoder structure of turbo codes. Source: https://spectrum.ieee.org/turbo-codes.

4.2.1.1 Turbo Codes

In 1993, two French engineers, C. Berrou and A. Glavieux, invented a new class of ECC codes, called Turbo codes [5]. The encoding structure of a Turbo code, as shown in Figure 4.2, contains two or more encoders and an interleaver to provide data redundancy for better error correction. The encoders are usually convolutional codes combined with a parallel cascade scheme. They perform the encoding calculations and generate the error correction parity bits (encoded data) for a given input. The two encoders have the same structure but different inputs. On the other hand, the pseudorandom interleaver permutes the bits in the message sequence. The permuted sequences are used as the input sequences of the second encoder. Now, let us see how the Turbo code error correction works.

Let us assume a message 01101 is to be encoded and transmitted. The message is passed directly to the transmitter, encoder 1, and the interleaver. The interleaver permutes the message and generates a permuted string 10011, which serves as the input of the encoder 2. The encoder 1 and encoder 2 perform the encoding calculation and generate the error correction parity bits, 10110 and 11100, respectively. By receiving the message and the parity bits, the transmitter concatenates (combines) the message with the encoder parity bits and transmits it to the network. Hence, the transmitter transmits the string 01101, 10110, 11100 to the network.

The receiver receives the transmitted string as an analog signal. It samples it and assigns an integer value to each received bit as a measure of the likeliness of the bit to be a 0 or a 1. The range could be from +7 to −7 or much bigger in real systems. In our case, +7 and −7 indicate that the bit is sure a 1 and a 0, respectively. Similarly, +5 and −5 indicate that the bit is likely 1 and 0, respectively. The least likeliness of a 1 is +1, while it is −1 for a 0. Our example described in Figure 4.2 shows that an error is inserted in the received message where a 1 is erroneously changed to 0. Hence, the receiver gives the likeness value of −2. The transmitter sends the likeness values corresponding to the message string to both the interleaver and the decoder 1. It also sends the likeness values corresponding to encoder 1 parity bits to the decoder 2, and that of the encoder 2 to the decoder 2. Each decoder receives the likeness values and evaluates its confidence in its actual value. The encoders exchange their confidence information until they conclude the final output, which is the original error-corrected and free message (Figure 4.2).

Although the MLD for Turbo codes is too difficult due to the parity-check matrices being constructed use of pseudorandom interleavers, Turbo codes have good structures so that it can use an efficient iterative decoding. The soft output of the decoder of the first component code is input to the decoder of other component code and vice versa until the final decoder estimate is obtained or the maximum iteration number is reached. This efficient suboptimal iterative decoding almost always converges to the result of MLD. Moreover, the best interleavers can permute the bits in the raw information sequence by using a pseudorandom manner, such that Turbo codes can have the property of random-like codes, to result in the best distance distribution.

Due to the above features, Turbo codes have excellent performance, especially at moderate bit error rate (BER) and for long block lengths. For any code rates and block lengths greater than 10^4 bits, the performance gap between Turbo codes using the iterative decoding algorithm and the Shannon limit can be within 1 dB when the decoding BERs are less than 10^{-5}. For instance, simulation results showed that the performance gap between a Turbo code using a random interleaver with a block length of 65,536 and code rate of 0.5, and the Shannon limit can be within 0.7 dB over the AWGN channel at the BER as low as 10^{-5}. Due to their excellent decoding performance, Turbo codes have been widely applied in various communication systems such as 3G and 4G. The IEEE 802.16 standards have adopted Turbo codes as their channel coding schemes. Moreover, the Consultative Committee for Space Data Systems (CCSDS) has listed Turbo codes as a proposed channel coding scheme for satellite communication and deep space communication [6].

4.2.1.2 LDPC Codes

In 1962, R. Gallager proposed a class of linear block codes with excellent performance, called LDPC codes [7], whose parity-check matrices have very low density (the number of nonzero elements is far less than the number of zero). However, due to their parity-check matrices being constructed pseudorandomly and the limitations of the hardware at that time, LDPC codes did not get much attention from coding researchers.

In 1981, R. Tanner gave a new interpretation of LDPC codes by introducing a graph, now called as Tanner graph [8], to simply represent LDPC codes. However, this important work was also ignored by coding researchers. A Tanner graph is composed of two types of nodes (variable nodes [VN] and check nodes [CN]) and edges which connect the nodes. Each VN in a Tanner graph represents each codeword symbol and corresponds to each column in the parity-check matrix of one LDPC code. Each CN in a Tanner graph represents each parity-check constraint of one LDPC code and corresponds to each row in the parity-check matrix. If a VN is connected to a CN by an edge, it represents that this VN is involved in the check constraint of the corresponding CN. The number of nonzero elements in each column of the parity-check matrix is called the column weight of this column or the degree of the corresponding VN. The number of nonzero elements in each row is called the row weight of this row or the degree of the corresponding CN. If one LDPC code has all column weights equal to a constant, and all row weights equal to a constant, it is called a regular code; otherwise, it is an irregular code. If one LDPC code has all column weights equal to a constant γ, and all row weights equal to a constant ρ, it is called a regular code; otherwise, it is an irregular code (Figure 4.3).

Until Turbo codes were discovered, many researchers noticed that Turbo codes could be viewed as a class of LDPC codes from a certain point of view, then LDPC codes ignored for more than 30 years were rediscovered and further generalized. With the development of graph and iterative decoding, the study of LDPC codes has made great advances in some key technologies:

- Constructions of structured LDPC codes
- Efficient encoding and decoding algorithms
- Hardware implementation

Many LDPC codes have been proposed that have excellent performance with linear encoding and decoding complexities. Moreover, the decoding algorithms

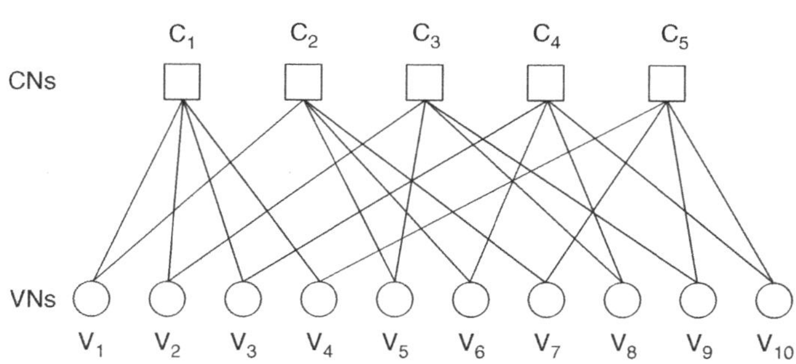

Figure 4.3 Parity-check matrix and Tanner graph of a (10,5) code.

of LDPC codes can achieve completely parallel computation in theory, which is convenient for hardware implementation. Analysis and simulation results have shown that long LDPC codes with implementable iterative decoding can achieve near-Shannon-limit performance. Simulation results showed that the decoding performance of one LDPC code of 0.5 rate and block length of 10^7 can approach the Shannon limit within 0.040 dB for the AWGN channel at the BER of 10^{-6}. Compared with Turbo code, LDPC code has some other advantages:

- It does not need the pseudorandom interleaver.
- Better error performance.
- Lower error floor.

The decoders of LDPC code are message-passing processes, where soft reliability messages are iteratively computed and passed between VNs and connected with CNs in the Tanner graph. The practical decoding computations for LDPC codes are decomposed into a larger number of smaller independent atomic units to achieve higher parallelism in hardware implementation. The computation in each VN is called VN update (see Figure 4.4), which calculates the messages that VN passes to each connected CN. The computation in each CN is called CN update, which calculates the messages the CN passes to each connected VN.

The sum–product algorithm (SPA) is the optimal iterative belief propagation (BP) decoding algorithm for binary LDPC codes, which has the fastest iteration convergence rate. However, the computation complexity of CN update of SPA is very high and requires look-up tables in actual hardware implementation. Thus, the decoder of SPA costs a lot of logical resources. The min-sum (MS) decoding algorithm is an approximate simplification algorithm of the SPA, which uses comparing and minimizing operations to replace complex hyperbolic tangent operations in CN updates (see Figure 4.5). Thus, the MS algorithm significantly reduces the computation complexity, and the decoder of the MS algorithm is easy to design and realize.

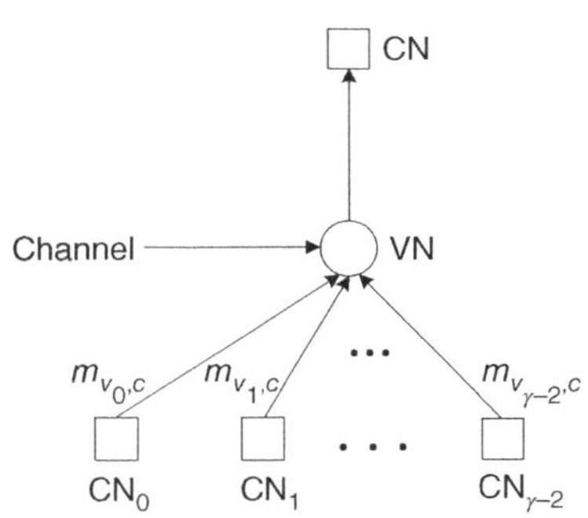

Figure 4.4 VN update diagram.

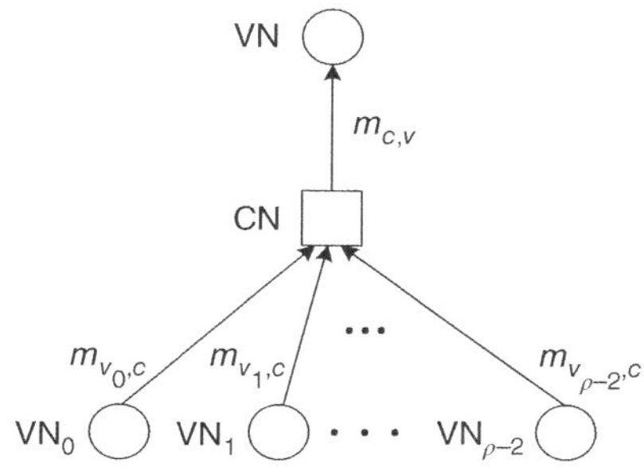

Figure 4.5 CN update diagram.

Although the LDPC code was rediscovered several years later than the Turbo code, LDPC codes have been widely applied in modern communication systems, e.g., DVB-S2, WiMax, Wi-Fi, and 10 G-base T communication protocols. LDPC codes have also been used in deep space communications, optical fiber communications, magnetic/light/holographic storages, and other areas. Moreover, due to their high computational efficiency, LDPC codes can meet the requirements of 5G on high throughput, low latency, and high hardware efficiency [9]. Hence, LDPC codes have been adopted as the standard channel coding for the data channels of the 5G eMBB scenario.

4.2.1.3 Polar Codes

In 2009, E. Arikan invented a class of channel capacity-achieving codes, called Polar codes [10]. Polar codes are the only coding scheme that was theoretically proved for the first time that it can achieve the Shannon limits of binary input symmetric discrete memoryless channels with the encoding complexity $O(N\log_2 N)$ and the decoding complexity $O(N\log_2 N)$, where $N = 2^m$ is the block length. The core of construction in the Polar code is the channel polarization effect. In the encoder, by adopting the coding method, each sub-channel presents different reliability. As the number of N increases, the ratio of good channels is

$$\left\lfloor \frac{\text{The number of sub-channels satisfing } I(W_i) > 1 - \delta}{N} \right\rfloor \to I(W)$$

and the ratio of bad channels is

$$\left\lfloor \frac{\text{The number of sub-channels satisfing } I(W_i) < \delta}{N} \right\rfloor \to 1 - I(W)$$

where $I(W_i)$ is the capacity of the ith polarized sub-channel, $\delta \in (0, 1)$. It has been verified by E. Ariken that

$$\delta \approx 2^{-\sqrt{N}}$$

Hence, when the block length N tends to infinity, good sub-channels will tend to the noiseless channels (channel capacity is close to 1), and the rest bad

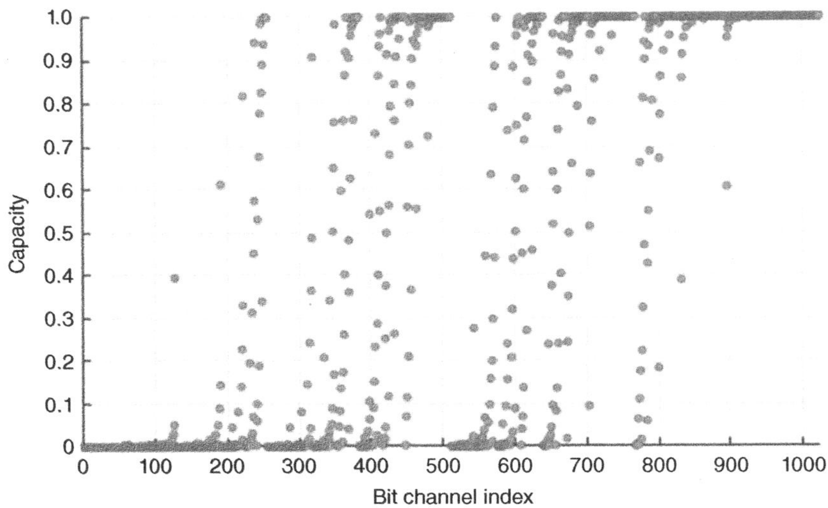

Figure 4.6 Capacity of polarized sub-channels with block length of 1024 [11]. Source: Bioglio, V., Condo, C. and Land, I., 2020/with permission of IEEE.

sub-channels will tend to the full noise channels (channel capacity is close to 0), i.e., the channel polarization phenomenon. Moreover, the ratio of noiseless sub-channels goes to the overall capacity of the channel as block length N tends to infinity, i.e., m tends to infinity. The encoding strategy of Polar code is to use all good channels to transmit useful messages and use bad channels to transmit known messages, which are called frozen bits or no messages so that it can achieve channel capacity and enjoy error-free transmission with the successive cancellation (SC) decoding (Figure 4.6).

However, the performance of Polar codes of finite block lengths with SC decoding is worse than Turbo codes and LDPC codes. Until the SC list (SCL) decoding and its efficiently improved algorithms have been successively proposed, Polar codes can obtain significant performance gain in short block lengths with the aid of cyclic redundancy check (CRC). Hence, although Polar codes may be less mature in practical applications compared with the existing two capacity-approaching codes, Polar codes have great application potential in next-generation mobile communication systems [11]. In 2017, Polar codes were adopted as the standard coding scheme for the control channels of the 5G eMBB scenario [12].

4.3 Security Codes

Security coding is a set of techniques used to protect data and communications from unauthorized access, interception, and modification. In 5G networks, security coding is used to protect sensitive information and ensure that communications are secure and private.

5G networks use various security measures including encryption, authentication, and access control to protect communications. Encryption involves converting data into a coded form that can only be read by authorized parties. Authentication involves verifying the identity of parties involved in a communication. Access control involves limiting access to network resources based on user credentials and other factors.

One of the main security features of 5G networks is end-to-end encryption. End-to-end encryption ensures that data is encrypted from the point of transmission to the point of reception, which prevents unauthorized parties from intercepting or modifying the data. This is achieved through the use of public key cryptography, which involves using a public key to encrypt data and a private key to decrypt it.

Another security feature of 5G networks is mutual authentication. Mutual authentication involves verifying the identities of both the user equipment (UE) and the network, which helps prevent unauthorized access to the network. This is achieved through the use of digital certificates, which are used to verify the identity of the UE and the network.

Access control is also an important security feature of 5G networks. Access control involves limiting access to network resources based on user credentials and other factors. This can be achieved through the use of access control lists (ACLs) and other access control mechanisms.

4.3.1 Public Key Infrastructure

Public key infrastructure (PKI) is a system for secure communication that relies on digital certificates and public key cryptography. It provides a way to verify the identity of a party in a communication and to ensure that the information exchanged between the parties is protected from eavesdropping and tampering [13, 14].

In a PKI system, each party has a public key and a private key. The private key is kept secret and used to decrypt messages that are encrypted with the corresponding public key. The public key, on the other hand, is freely distributed and is used by others to encrypt messages that can only be decrypted by the holder of the private key. An illustration of a simplified PKI is shown in Figure 4.7.

Public key cryptography has several advantages over traditional symmetric cryptography, where the same key is used for both encryption and decryption. First, it eliminates the need for a secure key distribution channel. In symmetric cryptography, the same key must be shared by both parties in advance through a secure channel, which can be difficult to achieve in practice. In contrast, with public key cryptography, the public key can be freely distributed, and there is no need for a secure channel for key distribution.

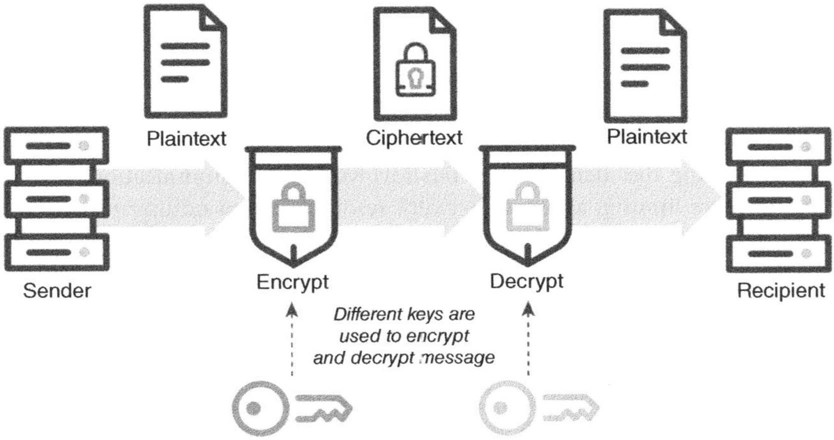

Figure 4.7 A simplified public key infrastructure (PKI). Source: https://www.sectigo.com/resource-library/what-is-pki-public-key-infrastructure.

Second, public key cryptography provides a way to authenticate the identity of the owner of a public key. Digital certificates, issued by trusted third parties, can be used to verify the authenticity of a public key. This allows users to trust the public key of a remote party, even if they have never communicated with that party before.

Public key cryptography is based on the mathematical properties of prime numbers and modular arithmetic. The security of the system depends on the difficulty of certain mathematical problems, such as factoring large integers, which are believed to be computationally infeasible for current computers. In the next section, we will discuss major existing cryptographic standards and algorithms.

4.3.2 Symmetric and Asymmetric Cryptography Concepts

Symmetric and asymmetric security codes are encryption techniques used to secure data transmission and communication between devices [15]. These algorithms use security keys to secure messages. The longer the key size, the better the security of the algorithm, as it becomes harder to brute force the secured message. Hence, the security of a cryptographic mechanism depends on the key size and the strength of its algorithm. Cryptos are categorized into symmetric and asymmetric key cryptos.

4.3.2.1 Symmetric Key Cryptography

Symmetric key cryptography uses the same secret key for both encryption and decryption of data, which is why it is also called secret-key cryptography. This key is typically shared between the sender and receiver of the data and must

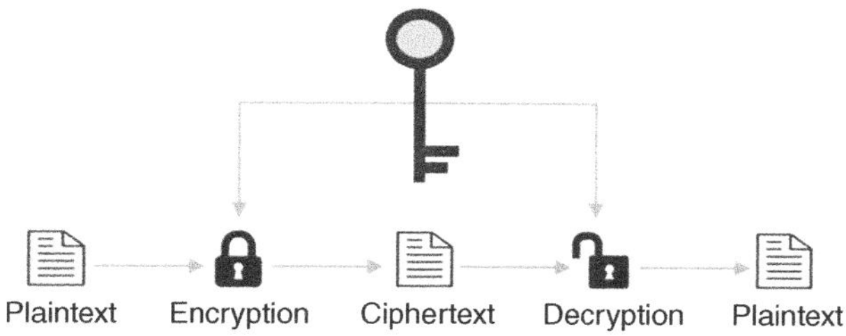

Figure 4.8 Symmetric key cryptography. Source: Sectigostore.com/https://sectigostore.com/blog/5-differences-between-symmetric-vs-asymmetric-encryption/.

be kept secret to prevent unauthorized access. The encryption and decryption process in symmetric key cryptography is computationally faster and more efficient than asymmetric key cryptography because the same key is used for both operations [16]. Some commonly used symmetric key encryption algorithms are the Advanced Encryption Standard (AES), ChaCha20, Serpent, and Triple DES (3DES). Other secret-key encryption algorithms include the Data Encryption Standard (DES – already broken), Twofish, Blowfish, RC4, RC5, RC6, and IDEA (Figure 4.8).

However, symmetric key cryptography has a significant drawback when it comes to secure key exchange. Since the same key is used for encryption and decryption, it needs to be shared between the sender and receiver. If an attacker intercepts the key during transmission, they can decrypt the data and compromise the security of the communication channel. Therefore, symmetric key cryptography is best suited for encrypting data that is transmitted over a secure channel, such as a local area network (LAN).

4.3.2.2 Asymmetric Key Cryptography

Asymmetric key cryptography, on the other hand, uses two mathematically related keys, a public key and a private key, for encryption and decryption of data. The public key is available to anyone, while the private key is kept secret. This technique is also called public key cryptography because the public key can be shared with anyone who wants to communicate with the private key owner. The encryption process uses the public key, while the decryption process uses the private key. Some commonly used asymmetric key encryption algorithms are RSA and elliptic curve cryptography (ECC) (Figure 4.9).

Asymmetric key cryptography is commonly used for authentication, digital signatures, and secure key exchange [17]. It allows the recipient of the data to verify the sender's identity and ensure the integrity of the message. This ensures that the message has not been tampered with during transmission and that it was indeed sent by the claimed sender.

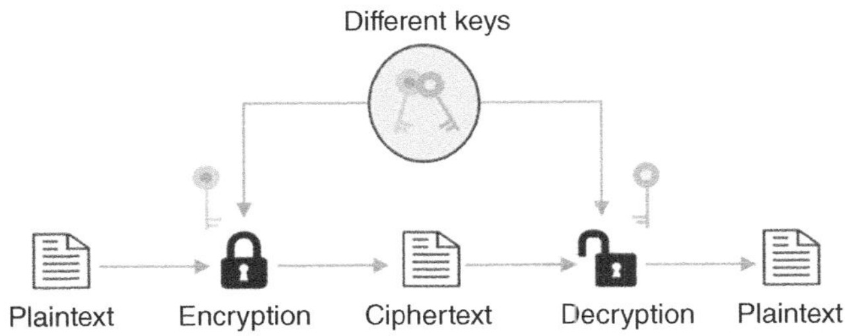

Figure 4.9 Asymmetric key cryptography. Source: Sectigostore.com/https://sectigostore.com/blog/5-differences-between-symmetric-vs-asymmetric-encryption/.

Digital signatures provide a way for the sender of a message to sign the message using their private key, which can then be verified using their public key. This proves the message's authenticity, as the signature can only be generated by the sender's private key. Digital signatures are used to provide non-repudiation, which means that the sender cannot deny sending the message [18]. Non-repudiation provides proof of the authenticity of a communication, preventing the sender from denying that they sent the message.

Non-repudiation is particularly important in wireless networks such as 5G because these networks are often used for critical communications and transactions, such as financial transactions and sensitive business information. Without non-repudiation, it would be difficult to verify the authenticity of these communications and transactions, which could lead to disputes or even fraud.

Asymmetric key cryptography is also used for secure key exchange between parties, which is essential for establishing a secure communication channel. One commonly used protocol for secure key exchange is the Diffie–Hellman key exchange, which allows two parties to establish a shared secret key over an insecure communication channel.

Asymmetric key cryptography also allows for the creation of digital certificates. In a 5G network, digital certificates are used to verify the identity of a user or device before allowing access to the network. For example, a user might present their digital certificate to a network access point, which can then use PKI to verify the authenticity of the certificate and grant access to the network. This provides a way to prevent unauthorized access to the network and ensures the communications are authentic and trustworthy.

In summary, symmetric key cryptography is fast and efficient for encrypting large amounts of data, while asymmetric key cryptography offers better security and is used for authentication, digital signatures, and secure key exchange. Both techniques have their strengths and weaknesses and are used in combination with modern cryptographic systems to provide a balance between efficiency and security.

4.3.3 Existing Algorithms and Standards

This section briefly describes existing AES, RSA, and ECC standards.

4.3.3.1 Advanced Encryption Standard

AES is a symmetric key encryption algorithm first adopted as a standard by the US government in 2001. It is widely used in applications that require high levels of security, such as military and government communications, as well as in commercial applications such as online banking and e-commerce transactions.

AES is a block cipher with a block size of 128 bits and the key length can be 128, 192, or 256 bits [19]. The algorithm consists of a series of mathematical operations performed on the data block and the key. The key is used to create a series of subkeys, which are then used in a series of rounds to transform the data block into ciphertext.

One of the strengths of AES is that it is resistant to attacks, such as brute-force attacks, that try to guess the encryption key by trying every possible key combination. This is because AES uses a key schedule that generates a large number of subkeys from the original key, making it more difficult to guess the correct key.

4.3.3.2 RSA Algorithm

RSA is an asymmetric key encryption algorithm developed by Ron Rivest, Adi Shamir, and Leonard Adleman in the 1970s. It is named after their last initials. RSA is widely used in applications such as secure messaging, digital signatures, and online transactions. It works by using a pair of keys: a public key and a private key [20]. The public key is used to encrypt messages, while the private key is used to decrypt them. The keys are mathematically related so that it is practically impossible to derive the private key from the public key.

To encrypt a message using RSA, the sender first obtains the recipient's public key (n, e, known as the public modulus and exponent, respectively). The message is then converted into a numerical value m, and the sender raises this value to the power of the public exponent e and takes the result modulo the public modulus n. The resulting value is the encrypted message called ciphertext c:

$$c = m^e \bmod n$$

To decrypt the ciphertext, the recipient uses their private key d, to raise the ciphertext to the power of the private exponent and takes the result modulo the modulus n:

$$m = c^d \bmod n$$

One of the strengths of RSA is that it provides a high level of security, even against brute-force attacks. However, it can be computationally intensive, particularly for large key sizes. Additionally, RSA is vulnerable to attacks that exploit the mathematical properties of the algorithm, such as attacks that factor the modulus n.

To mitigate these risks, it is important to use a sufficiently large key size, typically at least 2048 bits, and to implement additional security measures, such as key management and access control, to protect the private key. Despite its limitations, RSA remains one of the most widely used encryption algorithms in the world, and it is considered a highly secure and reliable encryption standard for protecting sensitive information.

4.3.3.3 Elliptic Curve Cryptography

ECC is a public key encryption algorithm based on the properties of elliptic curves over finite fields [21]. ECC is widely used in applications requiring strong security and efficient computation, such as mobile devices, smart cards, and the Internet of Things (IoT), with memory, power, or bandwidth limitations. It works by using a pair of keys: a public key and a private key. The public key is used to encrypt messages, while the private key is used to decrypt them. The security of ECC is based on the difficulty of their discrete logarithm problem. The keys are mathematically related (discrete logarithmic relationship) such that it is practically impossible to derive the private key from the public key. ECC is based on the properties of elliptic curves, which are curves defined by the equation

$$y^2 = x^3 + ax + b$$

where a and b are constants. The curve is defined over a finite field, a set of numbers that is limited to a specific range. ECC operations, such as point addition and scalar multiplication, are performed on the points that lie on the elliptic curve. The private key is a large unique or pseudorandom integer d. On the other hand, the public is a point on the curve Q, such that $=dG$, where G is the generator point of the curve.

To encrypt a message using ECC, consider the scenario where Alice, having a private key n_A and a public key $Q_A = n_A * G$, wants to send the message M to Bob whose private and public keys are n_B and $Q_B = n_B * G$, respectively. Alice computes the ciphertext C as the concatenation of the Alice's public key Q_A and the point $(M + n_A Q_B)$, i.e.,

$$C = \{\, Q_A, (M + n_A Q_B) \,\}$$

Bob receives the ciphertext and recovers the original message M by subtracting $n_B Q_A$ from the point $(M + n_A Q_B)$ in the ciphertext, i.e.,

$$M = (M + n_A Q_B) - n_B Q_A$$

One of the strengths of ECC is that it provides a high level of security with relatively small key sizes. ECC has a smaller key size compared to RSA at a similar level of security. It also has faster key and signature generation. This makes ECC particularly well suited for use in resource-constrained environments, such as mobile devices and the IoT. Additionally, ECC is resistant to many types of attacks, including brute-force attacks and attacks that exploit the mathematical properties of the algorithm.

4.4 Emerging 5G Security Design Acceleration

Future trends for 5G security codes will likely focus on using advanced technologies and methods to provide secure and reliable connectivity for users and devices. These trends will help address the challenges associated with securing large-scale, distributed networks and enable the development of new, innovative applications and services that can benefit users and organizations alike. In this section, we will cover key topics of blockchain, lightweight encryption algorithms, network codes, post-quantum cryptography, homomorphic encryption, and zero-knowledge proof.

4.4.1 Blockchain

Blockchain is a distributed ledger technology that enables secure and transparent recording and sharing of data across a network of computers. In a blockchain, each block of data is connected to the previous block using cryptographic techniques, creating a chain of blocks. The decentralized nature of blockchain technology means that no central authority controls the data, making it resistant to tampering and fraud.

Blockchain technology is particularly relevant in the context of 5G communications due to the vast amount of data generated and transmitted by these networks. 5G networks are expected to facilitate new applications and use cases, such as IoT devices, autonomous vehicles, and smart cities, which requiring secure and efficient data sharing, storage, and management [22]. Blockchain for 5G is illustrated in Figure 4.10.

One of the key benefits of blockchain technology is its ability to create a secure and transparent environment for data sharing and management. By creating a decentralized network of nodes that collectively maintain a tamper-proof record of transactions, blockchain technology can ensure that data is shared securely and transparently between different parties.

In the context of 5G networks, blockchain can be used to facilitate secure and efficient data sharing between different service providers, such as healthcare providers, financial institutions, and government agencies [23]. For example, a blockchain-based platform could be used to securely share patient data between

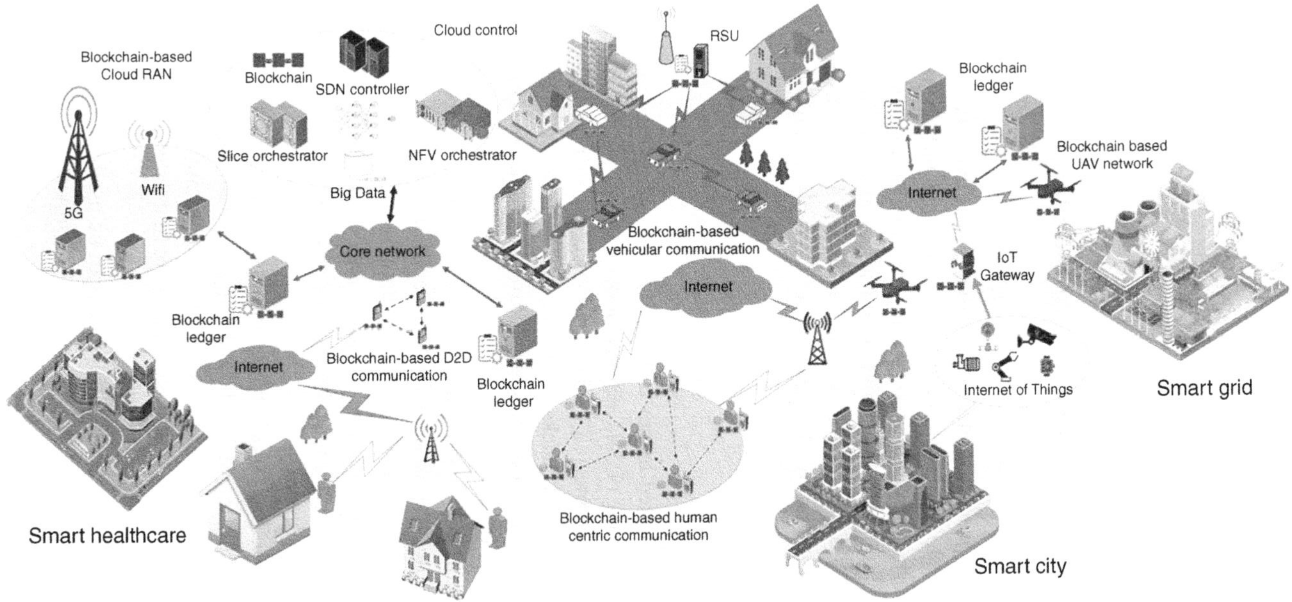

Figure 4.10 Blockchain for 5G [22]/with permission of Elsevier.

different healthcare providers, ensuring that sensitive information is kept private while still allowing for efficient data sharing.

One of the key benefits of blockchain technology is its ability to provide a tamper-proof and immutable record of data transactions. This means that once data is stored on the blockchain, it cannot be modified or deleted, ensuring the integrity and authenticity of the data. This is particularly important in the context of 5G networks, which will generate vast amounts of data that need to be securely stored and managed.

In addition to data security, blockchain technology can also facilitate interoperability between different service providers and networks [24]. By creating a decentralized network of nodes that can communicate and share data, blockchain technology can enable seamless data exchange and collaboration between different networks and service providers. This can be particularly important in the context of 5G networks, which will require interoperability between different devices and networks to deliver the full range of services and applications.

Furthermore, blockchain technology can also help to build trust between different parties in the 5G ecosystem. By providing a transparent and auditable record of transactions, blockchain can help to establish trust and accountability between different service providers, users, and other stakeholders. This can be particularly important in sensitive applications such as healthcare, where data privacy and security are critical.

Blockchain can be used to securely manage and verify user identities in a 5G network. By creating a secure, tamper-proof record of user identities on the blockchain, it is possible to prevent identity theft and fraud.

Blockchain can be used to facilitate secure and transparent payment and billing in a 5G network [25]. For example, blockchain-based smart contracts can be used to automatically initiate payments when certain conditions are met. Blockchain technology can also facilitate the development of new business models and revenue streams in 5G networks. By using blockchain-based smart contracts to facilitate payments and billing, service providers can create new revenue streams and enhance the efficiency of their operations.

Blockchain technology can also be used to enhance network security in 5G networks. By creating a secure and tamper-proof record of network activity, blockchain can help detect and prevent attacks and other security threats. For example, blockchain-based platforms could be used to monitor network activity and detect any suspicious behavior, allowing network operators to take appropriate action to mitigate the threat.

Overall, blockchain technology has the potential to address many of the key challenges facing 5G networks, including data security and privacy, interoperability, and trust. By providing a secure, transparent, and decentralized platform for data sharing and management, blockchain technology can enable the full potential of 5G networks, unlocking new opportunities for innovation and growth.

4.4.2 Lightweight Encryption Algorithms

Lightweight encryption algorithms are designed to provide strong security while minimizing the computational and memory resources required for their implementations. These algorithms are typically used in resource-constrained environments, such as embedded systems, wireless sensor networks, and IoT devices, where energy consumption, processing power, and memory capacity are limited.

There are several design principles that are commonly used in lightweight encryption algorithms. These include:

- **Simplicity:** The algorithm should be simple and easy to implement, with minimal computational and memory requirements.
- **Low Power Consumption:** The algorithm should be designed to minimize power consumption, which is particularly important for battery-powered devices.
- **Low Latency:** The algorithm should have low latency, which is the time required to encrypt or decrypt a message.
- **Low Memory Requirements:** The algorithm should have low memory requirements, which is particularly important for devices with limited memory capacity.
- **Resistance to Side-Channel Attacks:** The algorithm should resist side-channel attacks, which are attacks that exploit information leaked by the device during the encryption process, such as power consumption or electromagnetic radiation.

Key examples of lightweight encryption algorithms include:

- **SIMON and SPECK:** These are families of lightweight block ciphers developed by the US National Security Agency (NSA). They are designed to be highly efficient and have low memory requirements.
- **PRESENT:** This is a lightweight block cipher that was developed by researchers at the University of Leuven. It is designed to have low computational and memory requirements. It is widely used in applications such as RFID tags and smart cards.

We will discuss these key example algorithms in the following subsections.

4.4.2.1 SIMON and SPECK Algorithms

SIMON and SPECK are families of lightweight block ciphers designed by the US NSA in 2013 [26]. These ciphers are intended to provide strong security with a low memory footprint and efficient performance on a wide range of platforms, from microcontrollers to high-end processors.

Both SIMON and SPECK are based on the same design principles but differ in their key and block sizes. SIMON is a family of ciphers with block sizes of 32, 48, 64, 96, and 128 bits, while SPECK is a family of ciphers with block sizes of 64, 96, and 128 bits. Both families have key sizes equal to the block size, simplifying the key scheduling algorithm and reducing the amount of memory required.

The key feature of SIMON and SPECK is their use of the Feistel network, a type of structure used in block ciphers to increase the diffusion and confusion properties of the cipher. The Feistel network consists of multiple rounds, each applying a nonlinear function to half of the input data and then combining the results with the other half of the input data using a linear function. The SIMON and SPECK ciphers use a modified Feistel network with a different number of rounds and a key scheduling algorithm for each block size. Both SIMON and SPECK are shown in Figure 4.11.

SIMON and SPECK also use a combination of Exclusive OR (XOR) and bit rotation operations to increase the security of the cipher. These operations are efficient and require only a small amount of memory, making them well-suited for lightweight environments.

One of the advantages of SIMON and SPECK is their small memory footprint. The ciphers require only a small amount of memory to implement, which is important for resource-constrained environments such as embedded systems and IoT devices [27]. Additionally, SIMON and SPECK are highly efficient, with fast encryption and decryption speeds that make them suitable for applications such as wireless sensor networks and mobile devices.

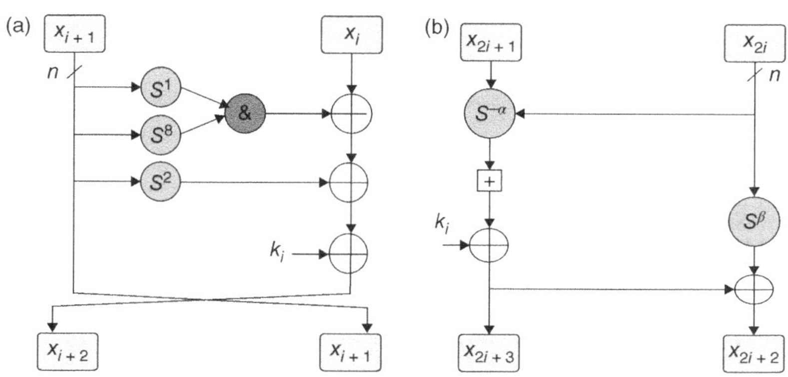

Figure 4.11 Feistel structure in block ciphers for (a) SIMON and (b) SPECK [27]/with permission of Springer Nature.

However, SIMON and SPECK have also been criticized for their design process, which was conducted secretly by the NSA and only made public after the ciphers were published. This has led to concerns about the trustworthiness of the ciphers and their potential susceptibility to backdoors or other vulnerabilities. Efforts have been made to study possible improvements for these two algorithms [28, 29].

4.4.2.2 PRESENT Algorithm

PRESENT is a lightweight block cipher designed in 2007 [30]. It is designed to have a small memory footprint and low power consumption, making it well suited for resource-constrained environments such as RFID tags, smart cards, and wireless sensor networks.

PRESENT has a block size of 64 bits and a key size of 80 bits. The key scheduling algorithm is simple and efficient and requires only a small amount of memory. The cipher uses a substitution–permutation network (SPN) structure with 31 rounds, each consisting of three main operations: a substitution layer, a permutation layer, and a key mixing operation. See Ref. [31] for the detailed description of the PRESENT encryption process (Figure 4.12).

The substitution layer in PRESENT utilizes an S-box with a 4-bit input and a 4-bit output. This S-box is engineered to resist both linear and differential cryptanalysis, using a blend of algebraic and combinatorial methods in its design. The permutation layer follows, consisting of a straightforward bit permutation that

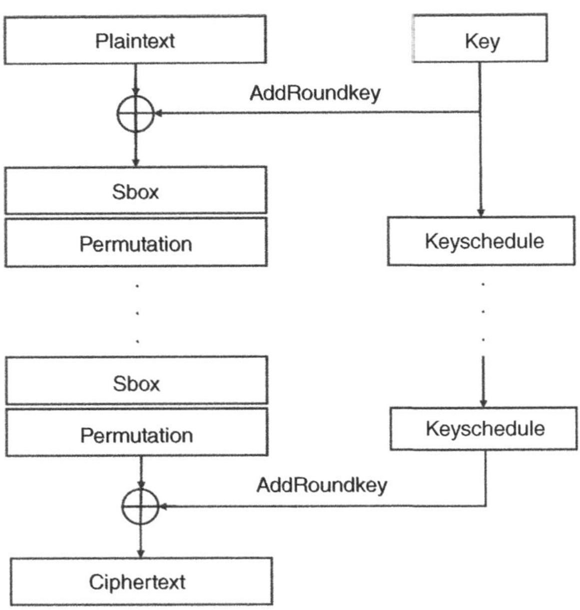

Figure 4.12 Encryption process of PRESENT block cipher [31]/MDPI/CCBY 4.0/Public domain.

rearranges the bits of the block based on a predefined pattern. Finally, the key mixing operation applies an XOR operation between the round key and the output from the permutation layer.

The S-box used in PRESENT is designed to be highly resistant to differential and linear cryptanalysis. It is based on the principle of "differential uniformity," which ensures that the output of the S-box is highly unpredictable even when the input is changed by a single bit. The S-box is constructed using a combination of an invertible function and a linear function, followed by an affine transformation.

The key schedule in PRESENT is simple and efficient, consisting of bitwise rotations, XOR operations, and S-box lookups. The main key is divided into 16 subkeys, each used in a specific round of the cipher. The subkeys are generated using a combination of the main key and a round constant that depends on the round number.

PRESENT is designed to have a strong avalanche effect, which means that a small change in the input or the key should result in a significant change in the output. This property is important for achieving confusion and diffusion, two important properties of a secure encryption algorithm.

PRESENT has undergone extensive cryptanalysis, including differential and linear cryptanalysis, algebraic cryptanalysis, and related key attacks. The cipher has been shown to be resistant to all known attacks and has a high-security margin.

Several variants of PRESENT have been proposed, including PRESENT-80 (with a 128-bit key size), PRESENT-128 (with a 128-bit block size and a 128-bit key size), and PRESENT-C (with a lightweight compression function). These variants are based on the same basic design principles as PRESENT but with different parameters and optimizations.

PRESENT is designed to be highly efficient in terms of area, power consumption, and latency. It can be implemented in hardware and software. It has been optimized for various platforms, including microcontrollers, smart cards, and wireless sensors [32].

PRESENT is known for its simplicity, efficiency, and security. The cipher is highly optimized for hardware implementations and can be implemented with as few as 1300 gates. Its low memory requirements and small code size make it well suited for applications with limited resources. The cipher has also been extensively studied by the cryptographic community and has been shown to be resistant to a wide range of attacks, including linear and differential cryptanalysis, and side-channel attacks [33].

However, PRESENT has also been criticized for its small key size, which is vulnerable to brute-force attacks. The 80-bit key size may not provide sufficient security against a determined attacker with access to large amounts of computing power. Additionally, the design process of PRESENT was not as transparent as some other ciphers, which has led to some concerns about its security and potential vulnerabilities.

4.4.2.3 GIFT Algorithm

GIFT (the Galois/Counter Mode Improved Family of Lightweight Block Ciphers) is a family of lightweight block ciphers that was first introduced in 2017 [34]. It is designed for resource-constrained environments, such as embedded systems and IoT devices, and is optimized for low power consumption and a small memory footprint.

GIFT is a block cipher, which means that it operates on fixed-size blocks of input data. GIFT comes in two versions: GIFT-64 and GIFT-128 [31], which have block sizes of 64 bits and 128 bits, respectively. Both versions use a 128-bit key and employ a Feistel network structure with 40 rounds, which means that the input block is split into two equal parts, and each part is processed separately in alternating rounds.

The round function in GIFT consists of four operations: a linear transformation, a bit permutation, a nonlinear substitution, and another bit permutation. The linear transformation and bit permutations are used to diffuse the input data, and the nonlinear substitution is used to add confusion to the output. The round function operates on both the left and right halves of the block, using a subkey that is generated by the key schedule. An illustration of the GIFT encryption process is shown in Figure 4.13.

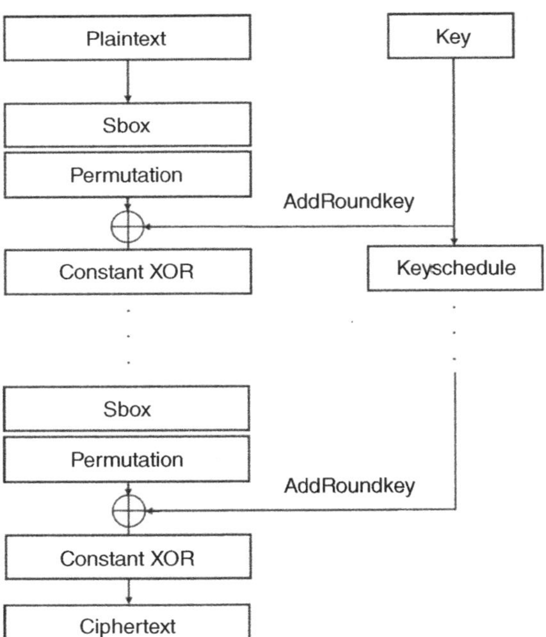

Figure 4.13 Encryption process of GIFT block cipher [31]/MDPI/CCBY 4.0/Public domain.

The key schedule in GIFT is simple and efficient. It requires only a small amount of memory. It generates a sequence of subkeys used in each round of the cipher. The key schedule is based on a key whitening technique, where the main key is XORed with a fixed round constant to generate the first subkey. The remaining subkeys are generated by applying a series of operations to the main key and the previous subkey.

One of the unique features of GIFT is its use of Galois/Counter Mode (GCM) for authenticated encryption. GCM combines a block cipher, such as GIFT, with a mode of operation that provides both confidentiality and integrity protection. This makes GIFT suitable for applications where data authenticity is also required, such as secure communication protocols.

GIFT has several advantages over other lightweight block ciphers. It has a small memory footprint and low power consumption, making it well suited for resource-constrained environments. It is optimized for use in GCM mode and provides both confidentiality and integrity for the encrypted data. It has undergone extensive cryptanalysis and is considered to be a secure cipher for lightweight applications. It also has good performance in both hardware and software implementations and is highly resistant to a wide range of attacks [33].

Overall, GIFT is a well-regarded lightweight cipher that provides strong security in resource-constrained environments. Its efficiency, resistance to attacks, and support for authenticated encryption make it a popular choice for use in embedded systems and IoT devices.

4.4.3 Network Codes

Network coding is a technique used in communication systems to improve the efficiency and reliability of data transmission. In traditional communication systems, data is transmitted in a series of packets. Each packet is sent individually and can contain data from one or more sources. However, when multiple packets are sent simultaneously, they can interfere with each other and lead to errors and delays.

Network coding is a technique that addresses this problem by combining multiple packets into a single transmission. This allows the receiver to decode the packets more efficiently improving the overall throughput and reducing the overall latency of the transmission. See Refs. [35, 36] for detailed descriptions of how network coding can achieve higher transmission efficiency over multicasting networks. Figure 4.14 illustrates how a source S transmits bits to receivers Y and Z. Node W produces a b_1 and b_2 exclusive-OR ($b_1 \oplus b_2$) bit and transmits this bit to X, which passes the bit to Y and Z. Node Z can decode b_1 from b_2 and, $b_1 \oplus b_2$ bit received. Similarly, Y can decide b_2 from receiving b_1 and $b_1 \oplus b_2$.

In 5G networks, network coding is used in combination with other techniques such as error coding and beamforming to improve the overall performance of the

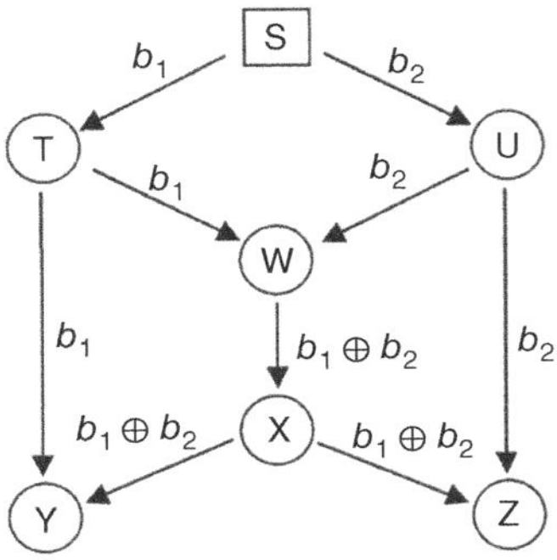

Figure 4.14 Multicast network illustrating network coding [36].

network. Network coding is primarily used in the downlink channel, where the base station transmits data to the UE. In this scenario, the base station can combine multiple packets of data from different sources, such as different users or different services, into a single transmission.

There are different types of network coding techniques used in 5G networks, including linear network coding and fountain coding. Linear network coding involves combining packets using linear equations, while fountain coding uses a probabilistic approach to encode packets.

Batched sparse (BATS) codes are a type of network coding scheme that combines fountain codes with random linear network coding [37]. Essentially, they are a way of encoding data into coded packets that can be distributed across multiple storage or network nodes, making the system more efficient and fault-tolerant. BATS codes consist of an outer code and an inner code, with the outer code being a matrix generalization of a fountain code and the inner code applying a linear transformation on each batch of data. The inner code is formed by performing linear network coding at the intermediate nodes with the constraint that only packets belonging to the same batch can be combined inside the network.

The key strengths of BATS codes are that they are designed to be robust against network topology and packet loss changes, which can significantly improve overall system performance. Additionally, they can operate with small finite fields, which reduce computational overhead and allow for more efficient encoding and decoding of data [38, 39].

Overall, BATS codes are a useful and versatile tool for designing efficient and reliable distributed storage and network systems, and they have applications in a wide range of fields and industries.

One key advantage of network coding is that it can be adapted to suit different transmission scenarios [40, 41]. For example, it can be optimized for different network topologies and different types of data, such as video or audio streams or online gaming. This flexibility makes it well suited for 5G networks, which require high-speed data transmission with low latency.

Another advantage of network coding is that it can improve the reliability of data transmission. By combining packets from multiple sources into a single transmission, network coding can reduce the impact of interference and other noise sources. This can improve the overall quality of the transmission and reduce the likelihood of errors or delays.

Another benefit of network coding in 5G networks is that it can reduce the power consumption of the network. By combining packets from multiple sources into a single transmission, network coding reduces the number of transmissions required to transmit the same amount of data. This, in turn, reduces the power consumption of the network.

Additionally, network coding can improve the security of data transmission in 5G networks. By combining packets from multiple sources into a single transmission, network coding makes it more difficult for an attacker to intercept and decipher the data. This is because the attacker would need to intercept and decode all packets in the transmission, which is more difficult than intercepting and decoding a single packet.

Furthermore, network coding can improve the quality of service (QoS) in 5G networks. QoS refers to the level of service that the network provides to different types of traffic. By combining packets from different sources into a single transmission, network coding can prioritize traffic and allocate network resources more efficiently, which can improve the QoS of the network.

In addition to the benefits of network coding, there are also some challenges associated with implementing this technique in 5G networks. One challenge is the increased computational complexity required to encode and decode network-coded transmissions. This can require more processing power and can increase the network's latency.

4.4.4 Post-Quantum Cryptography

Post-quantum cryptography is a type of cryptography designed to be secure against attacks by quantum computers. Quantum computers are a type of computer that uses quantum bits, or qubits, to perform calculations. Quantum computers are exponentially faster than classical computers for certain types of calculations, including those used in many cryptographic algorithms. This means that many encryption algorithms, such as RSA and ECC, currently used in 5G networks, may become vulnerable to attacks by quantum computers in the future [42, 43].

Post-quantum cryptography uses mathematical problems that are believed to be resistant to attacks by quantum computers. One of these mathematical problems is lattice-based cryptography. Lattice-based cryptography is a type of post-quantum cryptography that is based on the mathematical concept of lattices. A lattice is a set of points arranged in a regular pattern in a higher-dimensional space. Lattice-based cryptography uses the difficulty of certain computational problems related to lattices as the basis for its security.

In lattice-based cryptography, the security of the cryptographic algorithm relies on the hardness of the shortest vector problem (SVP) and the closest vector problem (CVP). SVP and CVP are two fundamental problems in lattice theory. SVP is the problem of finding the shortest nonzero vector in a lattice, while CVP is the problem of finding the lattice point closest to a given point [44].

To encrypt a message using a lattice-based cryptographic algorithm, the sender encodes the message as a point in a lattice. The sender then adds a random error vector to the encoded point and sends the resulting point to the receiver. The receiver then uses the lattice structure to recover the original message by finding the closest lattice point to the received point.

Lattice-based cryptography is believed to be resistant to attacks by quantum computers due to the difficulty of solving the SVP and CVP problems. Unlike many other cryptographic algorithms, the hardness of these problems does not depend on the factorization or discrete logarithm problems used in many classical cryptographic algorithms, which are known to be vulnerable to attacks by quantum computers.

Several lattice-based cryptographic algorithms have been proposed for post-quantum cryptography, including NTRU and Ring-LWE. In 2022, CRYSTALS-Kyber was selected by the National Institute of Standards and Technology (NIST) as the primary recommended algorithm for post-quantum (key-establishment) and CRYSTALS-Dilithium for digital signatures standardization [45]. The NIST Post-Quantum Cryptography Standardization Process has moved on to the fourth round with four alternate key-establishment candidate algorithms. Hence, BIKE, Classic McEliece, HQC, and SIKE will be further studied in the evaluation.

While lattice-based cryptography is promising as a post-quantum cryptographic solution, it is not without its challenges. One of the main challenges is the relatively high computational complexity and memory requirements of lattice-based cryptographic algorithms, which can make them less efficient than classical cryptographic algorithms. However, ongoing research is focused on improving the efficiency and practicality of lattice-based cryptographic algorithms to enable their use in a wide range of applications [46–48].

4.4.5 Homomorphic Encryption

Homomorphic encryption is a type of encryption that allows computations to be performed on encrypted data without first decrypting it. This is in contrast to traditional encryption, where data must be decrypted before it can be processed or analyzed. Homomorphic encryption allows computations on encrypted data

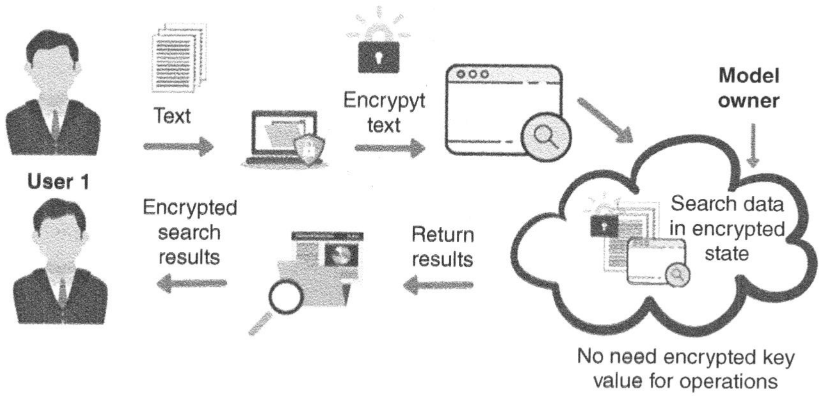

Figure 4.15 Homomorphic encryption illustration. Source: https://networksimulationtools.com/homomorphic-encryption-algorithm-projects.

without revealing the underlying data, providing an extra layer of privacy and security. An illustration of homomorphic encryption is shown in Figure 4.15.

In the context of 5G communications, homomorphic encryption can protect sensitive data while still allowing it to be processed and analyzed by applications and services [49]. For example, in a 5G network, homomorphic encryption can be used to encrypt user data, such as location information, without revealing the actual location to the network operator or any other third parties.

In addition to protecting user data, homomorphic encryption can also enable new applications and use cases in 5G. For example, homomorphic encryption can enable secure edge computing, where data is processed locally on edge devices rather than being sent back to a centralized data center. This can help reduce latency and improve performance while still ensuring that sensitive data remains protected. More information is discussed as follows.

(a) **Protecting User Privacy:**
As mentioned earlier, homomorphic encryption can help protect user privacy in 5G networks. For example, by encrypting user location data using homomorphic encryption, the network operator can still analyze the data to optimize network performance, without knowing the user's actual location [50]. This can help to alleviate privacy concerns for users who might be hesitant to share their location data with the network operator.

(b) **Secure Edge Computing:**
Edge computing is an important concept in 5G networks, as it allows data to be processed locally on edge devices rather than being sent back to a centralized data center. However, this raises concerns about data privacy and security since sensitive data might be processed on devices that are

not fully secured. Homomorphic encryption can help to address these concerns by allowing data to be processed locally on edge devices while remaining encrypted. This ensures that sensitive data remains protected, even if processed on an untrusted device.

(c) **Enabling New Applications and Use Cases:**

Homomorphic encryption is a relatively new technology, but it can potentially enable new applications and use cases in 5G networks. For example, homomorphic encryption can enable secure and privacy-preserving machine learning, where machine learning models can be trained on encrypted data without revealing the underlying data. This can be useful in applications, such as healthcare or finance, where data privacy is important.

(d) **Ongoing Research:**

Homomorphic encryption is still a developing field, and ongoing research is focused on developing more efficient and practical techniques. For example, efforts are being made to develop faster and more efficient homomorphic encryption algorithms, as well as techniques for optimizing computations on encrypted data. As these techniques become more practical and widely adopted, homomorphic encryption is likely to become an increasingly important tool for protecting sensitive data in 5G networks and other areas of information security.

Nonetheless, homomorphic encryption also presents some challenges, particularly around performance and complexity. Homomorphic encryption is typically slower and more computationally intensive than traditional encryption, which can hinder its adoption in some applications. There are also some limitations to what types of computations can be performed on homomorphically encrypted data.

Despite these challenges, homomorphic encryption is an important technology for protecting sensitive data in 5G communications and other areas of information security. Ongoing research is focused on developing more efficient and practical homomorphic encryption techniques that can be used in real-world applications.

4.4.6 Zero-Knowledge Proof

Zero-knowledge proof is a cryptographic technique that allows one party (the prover) to prove to another party (the verifier) that they know a secret without revealing any information about the secret itself [51]. In other words, zero-knowledge proof allows for authentication or verification without revealing

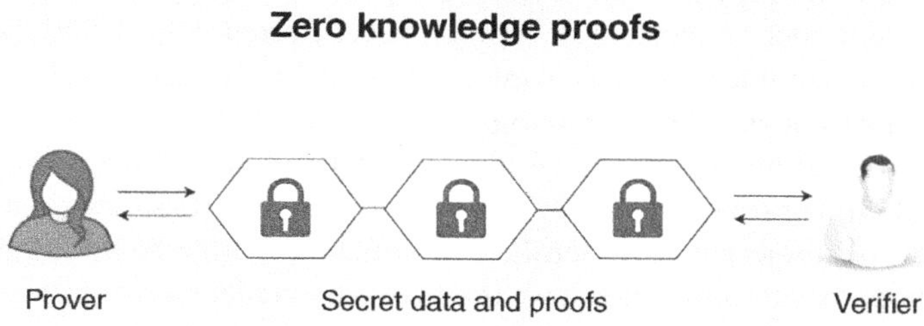

Figure 4.16 Illustration of zero-knowledge proof. Source: https://towardsdatascience.com/what-are-zero-knowledge-proofs-7ef6aab955fc.

any sensitive information. This is done through a series of interactive exchanges between the prover and verifier, where the prover provides some information proving they know the secret without revealing the secret itself. The verifier can then be convinced that the prover knows the secret without actually learning anything about the secret itself (Figure 4.16).

Several types of zero-knowledge proof exist, including interactive and non-interactive [23]. Interactive zero-knowledge proofs require the prover and verifier to engage in a series of back-and-forth exchanges, while non-interactive zero-knowledge proofs do not require any interaction between the prover and verifier. Non-interactive zero-knowledge proofs are typically more efficient but require more complex cryptography. Here, we will discuss some of the applications and challenges of zero-knowledge proof for 5G networks.

(a) **Applications in Access Control:**
One of the key applications of zero-knowledge proof in 5G networks is access control. In a 5G network, users may need to prove their identity to the network operator to access certain resources or services. With zero-knowledge proof, users can prove their identity without revealing any sensitive information, such as their name or other personal details. This can help to protect user privacy and prevent unauthorized access to the network [52].

(b) **Applications in Authentication:**
Zero-knowledge proof can also be used for authentication, allowing users to prove their identity without revealing sensitive information [53]. For example, users may need to authenticate themselves to a 5G network to access certain

resources or services. With zero-knowledge proof, the user can prove their identity to the network without revealing any sensitive information about themselves. Another application of zero-knowledge proof is for healthcare, where access to confidential medical information can be protected [54].

(c) **Applications in Secure Data Sharing:**
Zero-knowledge proof can also be used for secure data sharing and collaboration [55]. For example, in a 5G network, multiple parties may need to collaborate on a project or share sensitive information. With zero-knowledge proof, each party can prove they have the necessary credentials or information without actually revealing any sensitive information to the other parties. This can help ensure that sensitive information remains protected while still enabling collaboration and data sharing.

(d) **Challenges of Zero-Knowledge Proof:**
One of the main challenges of zero-knowledge proof is that it can be computationally expensive, particularly for interactive zero-knowledge proofs. In addition, zero-knowledge proof requires both parties to be actively involved in the authentication or verification process, which can be challenging in some situations.

Overall, zero-knowledge proof is an important technology for protecting sensitive information and maintaining privacy in 5G networks and other areas of information security. Its applications in access control, authentication, and secure data sharing make it a powerful tool for protecting user privacy and preventing unauthorized access to sensitive data. However, its implementation requires specialized knowledge and careful consideration of the specific use case. Ongoing research is focused on improving its efficiency and practicality for real-world scenarios.

References

1 Shannon, C.E. (1948). A mathematical theory of communication. *The Bell System Technical Journal* 27 (3): 379–423.

2 Simonoski, O., Marina, N., and Dimoska, N. (2021). Coding schemes in 5G networks for error-free communication. *Journal of Emerging Computer Technologies* 1 (2): 44–48.

3 Arora, K., Singh, J., and Randhawa, Y.S. (2020). A survey on channel coding techniques for 5G wireless networks. *Telecommunication Systems* 73: 637–663.

4 Cheng, S. (2019). Comparative study on 5G communication channel coding technology. *3rd International Conference on Mechatronics Engineering and Information Technology (ICMEIT 2019)*, 74–77. Atlantis Press, 29–30 March 2019, Dalian China.

5 Berrou, C., Glavieux, A., and Thitimajshima, P. (1993). Near Shannon limit error-correcting coding and decoding: turbo-codes. 1. *Proceedings of ICC'93-IEEE International Conference on Communications*, Vol. 2, pp. 1064–1070. 23–26 May 1993, Geneva Switzerland. IEEE.

6 The Consultative Committee for Space Data Systems (CCSDS). TM Synchronization and Channel Coding (131.0-B-1 Blue Book), Consultative Committee for Space Data Systems, September 2003.

7 Gallager, R. (1962). Low-density parity-check codes. *IRE Transactions on Information Theory* 8 (1): 21–28.

8 Tanner, R. (1981). A recursive approach to low complexity codes. *IEEE Transactions on Information Theory* 27 (5): 533–547.

9 Chen, J.I.Z. (2020). 5G systems with low density parity check based channel coding for enhanced mobile broadband scheme. *IRO Journal on Sustainable Wireless Systems* 2 (1): 42–49.

10 Arikan, E. (2009). Channel polarization: a method for constructing capacity-achieving codes for symmetric binary-input memoryless channels. *IEEE Transactions on information Theory* 55 (7): 3051–3073.

11 Bioglio, V., Condo, C., and Land, I. (2020). Design of polar codes in 5G new radio. *IEEE Communications Surveys & Tutorials* 23 (1): 29–40.

12 Hashemi, S.A., Condo, C., Ercan, F., and Gross, W.J. (2017). On the performance of polar codes for 5G eMBB control channel. *2017 51st Asilomar Conference on Signals, Systems, and Computers*, pp. 1764–1768. 6–9 November 2016, Pacific Grove CA USA. IEEE.

13 Benantar, M. (2001). The Internet public key infrastructure. *IBM Systems Journal* 40 (3): 648–665.

14 Dankers, J., Garefalakis, T., Schaffelhofer, R., and Wright, T. (2002). Public key infrastructure in mobile systems. *Electronics & Communication Engineering Journal* 14 (5): 180–190.

15 Al-Shabi, M.A. (2019). A survey on symmetric and asymmetric cryptography algorithms in information security. *International Journal of Scientific and Research Publications (IJSRP)* 9 (3): 576–589.

16 Maqsood, F., Ahmed, M., Ali, M.M., and Shah, M.A. (2017). Cryptography: a comparative analysis for modern techniques. *International Journal of Advanced Computer Science and Applications* 8 (6): 442–448.

17 Khan, A.G., Basharat, S., and Riaz, M.U. (2018). Analysis of asymmetric cryptography in information security based on computational study to ensure confidentiality during information exchange. *International Journal of Scientific Engineering and Research* 9 (11): 992–999.

18 Zhou, J. and Lam, K.Y. (1999). Securing digital signatures for non-repudiation. *Computer Communications* 22 (8): 710–716.

19 Abdullah, A.M. (2017). Advanced encryption standard (AES) algorithm to encrypt and decrypt data. *Cryptography and Network Security* 16: 1–12.

20 Milanov, E. (2009). The RSA algorithm. *RSA Laboratories* White Paper 1–11, 3 June 2009.

21 Sanka, A.I., Irfan, M., Huang, I., and Cheung, R.C. (2021). A survey of breakthrough in blockchain technology: adoptions, applications, challenges and future research. *Computer Communications* 169: 179–201.

22 Nguyen, D.C., Pathirana, P.N., Ding, M., and Seneviratne, A. (2020). Blockchain for 5G and beyond networks: a state of the art survey. *Journal of Network and Computer Applications* 166: 102693.

23 Dhar Dwivedi, A., Singh, R., Kaushik, K. et al. (2021). Blockchain and artificial intelligence for 5G-enabled Internet of Things: challenges, opportunities, and solutions. *Transactions on Emerging Telecommunications Technologies* 35: e4329.

24 Praveen, G., Chamola, V., Hassija, V., and Kumar, N. (2020). Blockchain for 5G: a prelude to future telecommunication. *IEEE Network* 34 (6): 106–113.

25 Maksymyuk, T., Gazda, J., Han, L., and Jo, M. (2019). Blockchain-based intelligent network management for 5G and beyond. *2019 3rd International Conference on Advanced Information and Communications Technologies (AICT)*, pp. 36–39. 2–6 July 2019, Lviv Ukraine. IEEE.

26 Beaulieu, R., Shors D., Smith J., Treatman-Clark, S., Weeks, B., Wingers, L., 2015. The SIMON and SPECK families of lightweight block cipers. *Proceedings of the 52nd ACM/EDAC/IEEE Design Automation Conference (DAC'15)*, 8–12 June 2015, San Francisco California USA, Article No.: 175, pp. 1–6.

27 Jangra, M. and Singh, B. (2022). Performance evaluation of SIMON and SPECK block ciphers to secure IoT-enabled smart cities. *Advanced Computing and Intelligent Technologies: Proceedings of ICACIT 2022*, pp. 451–461. 12–13 March 2022, Chennai India. Singapore: Springer Nature Singapore.

28 Alkhzaimi, H.A. and Lauridsen, M.M., 2013. Cryptanalysis of the SIMON family of block ciphers. *Cryptology ePrint Archive*, Paper 2013/543. Corpus ID: 8595056, pp. 1–26.

29 Dehnavi, S.M. (2018). Further observations on SIMON and SPECK block cipher families. *Cryptography* 3 (1): 1.

30 Bogdanov, A., Knudsen, L.R., Leander, G. et al. (2007). PRESENT: an ultra-lightweight block cipher. *Cryptographic Hardware and Embedded Systems-CHES 2007: 9th International Workshop,* Vienna, Austria (10–13 September 2007). Proceedings 9, 450–466. Berlin Heidelberg: Springer.

31 Jang, K., Song, G., Kim, H. et al. (2021). Efficient implementation of PRESENT and GIFT on quantum computers. *Applied Sciences* 11 (11): 4776.

32 Kim, H., Eum, S., Lee, W.K. et al. (2022). Secure and robust Internet of Things with high-speed implementation of PRESENT and GIFT block ciphers on GPU. *Applied Sciences* 12 (20): 10192.

33 Luo, H., Chen, W., Ming, X., and Wu, Y. (2021). General differential fault attack on PRESENT and GIFT cipher with nibble. *IEEE Access* 9: 37697–37706.

34 Banik, S., Pandey, S.K., Peyrin, T. et al. (2017). GIFT: a small present: towards reaching the limit of lightweight encryption. *Cryptographic Hardware and Embedded Systems–CHES 2017: 19th International Conference*, Taipei, Taiwan (25–28 September 2017), *Proceedings*, 321–345. Springer International Publishing.

35 Ahlswede, R., Cai, N., Li, S.Y., and Yeung, R.W. (2000). Network information flow. *IEEE Transactions on Information Theory* 46 (4): 1204–1216.

36 Yeung, R.W. (2006). *Network Coding Theory.* Now Publishers Inc.

37 Yang, S. and Yeung, R.W. (2014). Batched sparse codes. *IEEE Transactions on Information Theory* 60 (9): 5322–5346.

38 Yang, S., Yeung, R.W., Cheung, J.H., and Yin, H.H. (2014). BATS: network coding in action. *2014 52nd Annual Allerton Conference on Communication, Control, and Computing (Allerton)*, pp. 1204–1211. 1–3 October 2014, Monticello Illinois USA. IEEE.

39 Wang, S., Zhou, Q., Yang, S. et al. (2022). Wireless communication strategy with BATS codes for butterfly network. *Journal of Physics: Conference Series* 2218 (1): 012003.

40 Fragouli, C. and Soljanin, E. (2006). Information flow decomposition for network coding. *IEEE Transactions on Information Theory* 52 (3): 829–848.

41 Koetter, R. and Médard, M. (2003). An algebraic approach to network coding. *IEEE/ACM Transactions on Networking* 11 (5): 782–795.

42 Clancy, T.C., McGwier, R.W., and Chen, L. (2019). TUTORIAL: post-quantum cryptography and 5G security. *WiSec'19: ACM Conference on Security and Privacy in Wireless and Mobile Networks*. p. 285. 15–17 May 2019, Miami Florida USA.

43 Mattsson, J.P., Smeets, B., and Thormarker, E. (2021). Quantum technology and its impact on security in mobile networks. *Ericsson Technology Review* 2021 (12): 2–12.

44 Asif, R. (2021). Post-quantum cryptosystems for Internet-of-Things: a survey on lattice-based algorithms. *IoT* 2 (1): 71–91.

45 Alagic, G., Apon, D., Cooper, D. et al. (2022). Status report on the third round of the NIST post-quantum cryptography standardization process. US Department of Commerce, NIST.

46 Beri, R., Kaur, R., and Kaur, K. (2021). A contemporary study on quantum-computing security mechanisms in 5G networks. *Turkish Journal of Computer and Mathematics Education (TURCOMAT)* 12 (2): 450–455.

47 Partala, J. (2021). Post-quantum cryptography in 6G. *6G Mobile Wireless Networks* 431–448, Springer 2021.

48 Zeydan, E., Turk, Y., Aksoy, B., and Ozturk, S.B. (2022). Recent advances in post-quantum cryptography for networks: a survey. *2022 Seventh International Conference on Mobile and Secure Services (MobiSecServ)*, pp. 1–8. IEEE. 26–27 February 2022, Virtual Conference.

49 Sanon, S.P., Reddy, R., Lipps, C., and Schotten, H.D. (2023). Secure federated learning: an evaluation of homomorphic encrypted network traffic prediction. *2023 IEEE 20th Consumer Communications & Networking Conference (CCNC)*, pp. 1–6. 8–11 January 2023, Las Vegas Nevada USA. IEEE.

50 Yan, Z., Qian, X., Liu, S., and Deng, R. (2022). Privacy protection in 5G positioning and location-based services based on SGX. *ACM Transactions on Sensor Networks (TOSN)* 18 (3): 1–19.

51 Goldwasser, S., Micali, S., and Rackoff, C. (1985). The knowledge complexity of interactive proof- systems. *Proceedings of the Seventeenth Annual ACM Symposium on Theory of Computing (STOC '85)*, pp. 291–304, 6–8 May 1985, Providence Rhode Island USA. ACM.

52 Boubakri, W., Abdallah, W., and Boudriga, N. (2017). Access control in 5G communication networks using simple PKI certificates. *2017 13th International Wireless Communications and Mobile Computing Conference (IWCMC)*, pp. 2092–2097. 26–30 June 2017, Valencia Spain. IEEE.

53 Pathak, A., Patil, T., Pawar, S., Raut, P., Khairnar, S. and Gite, D., 2021. Bibliometric survey on zero-knowledge proof for authentication. Library Philosophy and Practice (e-journal) (LPP) ISSN 1522-0222, 30 January 2021. libphilprac/5046), pp. 1–26.

54 Gaba, G.S., Hedabou, M., Kumar, P. et al. (2022). Zero knowledge proofs based authenticated key agreement protocol for sustainable healthcare. *Sustainable Cities and Society* 80: 103766.

55 Tao, Y., Jiang, Y., Lin, P., Tsukada, M., and Esaki, H. (2023). zk-PoT: zero-knowledge proof of traffic for privacy enabled cooperative perception. *2023 IEEE 20th Consumer Communications & Networking Conference (CCNC)*, pp. 261–268, 8–11 January 2023, Las Vegas Nevada USA. IEEE.

Part III

Architecture

5

Hardware Architecture

This chapter revisits the 5G development and the related wireless technologies discussed in Chapter 2, focusing on strategies to optimize 5G performance from a hardware perspective.

5.1 Development Timeline

The International Telecommunication Union (ITU) establishes global standards for wireless mobile telecommunications technology with the goal of connecting people worldwide [1]. As part of this effort, ITU defined the framework for International Mobile Telecommunications (IMT) for 2020 and beyond, highlighting three key use cases: (i) enhanced mobile broadband (eMBB), (ii) ultra-reliable and low-latency communications (URLLC), and (iii) massive machine-type communications (mMTC) [2].

To meet ITU's requirements, the 3rd Generation Partnership Project (3GPP) develops technical specifications for the telecommunications industry [3]. Previously, 3GPP produced mobile communication standards for 2G, 3G, and LTE (4G). The organization began working on 5G standardization in 2015.

In 2018, 3GPP released its first comprehensive 5G standard, known as Release 15, which covered important aspects of 5G New Radio (NR), the 5G System (Phase 1), massive MTC, the Internet of Things (IoT), vehicle-to-everything services, further LTE enhancements, and mobile communication systems. This release also addressed the integration of non-standalone 5G systems with existing LTE networks. Release 16, issued in 2019, introduced further improvements to 5G and LTE, including updates to multiple-input and multiple-output (MIMO), NR for unlicensed spectrum, power efficiency, and vertical expansion.

Release 17 builds upon Release 16 by introducing enhancements in areas such as new use cases, industry verticals, and broader connectivity. Currently, 3GPP is working on Release 18, which focuses on 5G Advanced, incorporating

Edge Computing Acceleration: From 5G to 6G and Beyond, First Edition.
Patrick Hung, Hongwei Kan, and Greg Knopf.

new technologies and adding intelligence to mobile networks, such as machine learning. Figure 5.1 illustrates the timeline of major documents from both ITU and 3GPP.

The new release introduces significant advancements in the integration of artificial intelligence (AI) and machine learning (ML) for network optimization, enabling more intelligent and adaptive network operations. These technologies empower networks to dynamically adjust to real-time conditions, improving efficiency, reducing latency, and enhancing overall performance. This smarter approach to network management leads to better resource allocation, predictive maintenance, and automated troubleshooting, creating a more resilient and responsive 5G infrastructure.

Another key highlight of the release is the enhanced support for extended reality (XR) applications, which include augmented reality (AR) and virtual reality (VR). These improvements enable more immersive, interactive, and seamless experiences, paving the way for breakthroughs in fields such as entertainment, education, healthcare, and industrial training. The increased bandwidth and lower latency of 5G, combined with AI-driven optimizations, ensure that XR applications can operate smoothly even under demanding conditions.

Additionally, Release 18 expands the reach and performance of 5G networks in rural and remote areas, addressing the digital divide by ensuring that underserved regions can benefit from the latest mobile technologies. This broader connectivity enhances access to essential services such as telemedicine, remote education, and IoT-driven agriculture, while also boosting economic development in isolated communities. By extending 5G's capabilities to these areas, Release 18 helps realize the vision of truly global, inclusive connectivity.

5.2 Operating Spectrum

The 5G standard categorizes its operating spectrum into two distinct frequency ranges: Frequency Range 1 (FR1), which spans from 450 MHz to 7125 MHz, and Frequency Range 2 (FR2), which covers the higher spectrum from 24.25 GHz to 52.6 GHz, also known as millimeter-wave (mmWave) transmission. These ranges are defined in 3GPP TS38.101, which specifies the requirements for User Equipment (UE) radio transmission and reception [4, 5].

In addition to defining the frequency ranges, 3GPP TS18.101 outlines the specific operating bands within each range. Table 5.1, extracted from the latest version of TS18.101 [4], lists the operating bands for FR1, while Table 5.2 covers the FR2 bands [5]. The operating bands in these frequency ranges can support different duplexing techniques. Some bands utilize frequency division duplex (FDD), where two separate frequency bands are allocated for transmission and

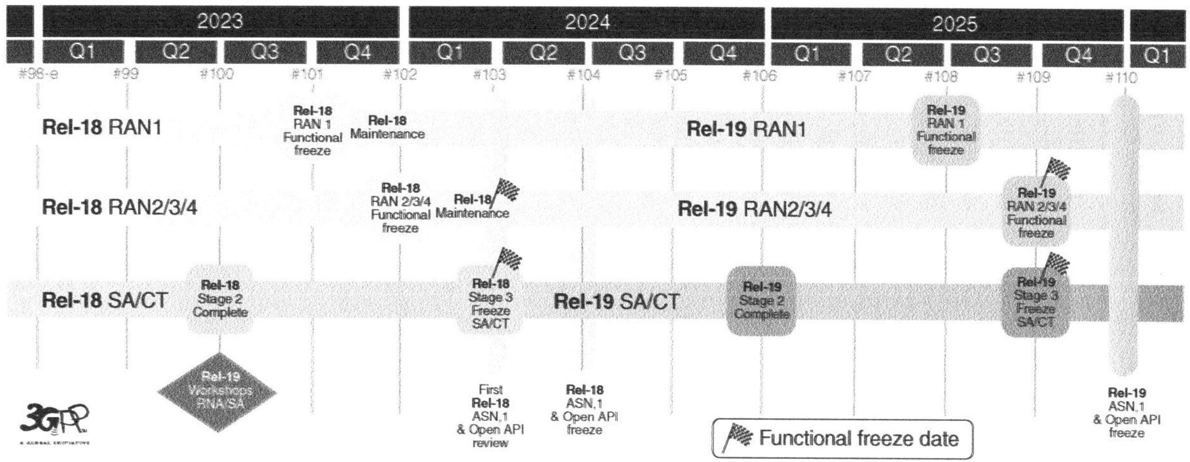

Figure 5.1 The 5G timeline: ITU-R and 3GPP releases from 5G network and 3GPP.

Table 5.1 NR operating bands in FR1.

NR operating band	Uplink (UL) operating band BS receive/UE transmit F_{UL_low} – F_{UL_high}	Downlink (DL) operating band BS transmit/UE receive F_{DL_low} – F_{DL_high}	Duplex mode
n1	1920 MHz – 1980 MHz	2110 MHz – 2170 MHz	FDD
n2	1850 MHz – 1910 MHz	1930 MHz – 1990 MHz	FDD
n3	1710 MHz – 1785 MHz	1805 MHz – 1880 MHz	FDD
n5	824 MHz – 849 MHz	869 MHz – 894 MHz	FDD
n7	2500 MHz – 2570 MHz	2620 MHz – 2690 MHz	FDD
n8	880 MHz – 915 MHz	925 MHz – 960 MHz	FDD
n12	699 MHz – 716 MHz	729 MHz – 746 MHz	FDD
n13	777 MHz – 787 MHz	746 MHz – 756 MHz	FDD
n14	788 MHz – 798 MHz	758 MHz – 768 MHz	FDD
n18	815 MHz – 830 MHz	860 MHz – 875 MHz	FDD
n20	832 MHz – 862 MHz	791 MHz – 821 MHz	FDD
n24	1626.5 MHz – 1660.5 MHz	1525 MHz – 1559 MHz	FDD
n25	1850 MHz – 1915 MHz	1930 MHz – 1995 MHz	FDD
n26	814 MHz – 849 MHz	859 MHz – 894 MHz	FDD
n28	703 MHz – 748 MHz	758 MHz – 803 MHz	FDD
n29	N/A	717 MHz – 728 MHz	SDL
n30	2305 MHz – 2315 MHz	2350 MHz – 2360 MHz	FDD
n34	2010 MHz – 2025 MHz	2010 MHz – 2025 MHz	TDD
n38	2570 MHz – 2620 MHz	2570 MHz – 2620 MHz	TDD
n39	1880 MHz – 1920 MHz	1880 MHz – 1920 MHz	TDD
n40	2300 MHz – 2400 MHz	2300 MHz – 2400 MHz	TDD
n41	2496 MHz – 2690 MHz	2496 MHz – 2690 MHz	TDD
n46	5150 MHz – 5925 MHz	5150 MHz – 5925 MHz	TDD
n47	5855 MHz – 5925 MHz	5855 MHz – 5925 MHz	TDD
n48	3550 MHz – 3700 MHz	3550 MHz – 3700 MHz	TDD
n50	1432 MHz – 1517 MHz	1432 MHz – 1517 MHz	TDD
n51	1427 MHz – 1432 MHz	1427 MHz – 1432 MHz	TDD
n53	2483.5 MHz – 2495 MHz	2483.5 MHz – 2495 MHz	TDD
n65	1920 MHz – 2010 MHz	2110 MHz – 2200 MHz	FDD
n66	1710 MHz – 1780 MHz	2110 MHz – 2200 MHz	FDD
n67	N/A	738 MHz – 758 MHz	SDL

Table 5.1 (Continued)

NR operating band	Uplink (UL) operating band BS receive/UE transmit $F_{UL\_low} - F_{UL\_high}$	Downlink (DL) operating band BS transmit/UE receive $F_{DL\_low} - F_{DL\_high}$	Duplex mode
n70	1695 MHz – 1710 MHz	1995 MHz – 2020 MHz	FDD
n71	663 MHz – 698 MHz	617 MHz – 652 MHz	FDD
n74	1427 MHz – 1470 MHz	1475 MHz – 1518 MHz	FDD
n75	N/A	1432 MHz – 1517 MHz	SDL
n76	N/A	1427 MHz – 1432 MHz	SDL
n77	3300 MHz – 4200 MHz	3300 MHz – 4200 MHz	TDD
n78	3300 MHz – 3800 MHz	3300 MHz – 3800 MHz	TDD
n79	4400 MHz – 5000 MHz	4400 MHz – 5000 MHz	TDD
n80	1710 MHz – 1785 MHz	N/A	SUL
n81	880 MHz – 915 MHz	N/A	SUL
n82	832 MHz – 862 MHz	N/A	SUL
n83	703 MHz – 748 MHz	N/A	SUL
n84	1920 MHz – 1980 MHz	N/A	SUL
n85	698 MHz – 716 MHz	728 MHz – 746 MHz	FDD
n86	1710 MHz – 1780 MHz	N/A	SUL
n89	824 MHz – 849 MHz	N/A	SUL
n90	2496 MHz – 2690 MHz	2496 MHz – 2690 MHz	TDD
n91	832 MHz – 862 MHz	1427 MHz – 1432 MHz	FDD
n92	832 MHz – 862 MHz	1432 MHz – 1517 MHz	FDD
n93	880 MHz – 915 MHz	1427 MHz – 1432 MHz	FDD
n94	880 MHz – 915 MHz	1432 MHz – 1517 MHz	FDD
n95	2010 MHz – 2025 MHz	N/A	SUL
n96	5925 MHz – 7125 MHz	5925 MHz – 7125 MHz	TDD
n97	2300 MHz – 2400 MHz	N/A	SUL
n98	1880 MHz – 1920 MHz	N/A	SUL
n99	1626.5 MHz – 1660.5 MHz	N/A	SUL
n100	874.4 MHz – 880 MHz	919.4 MHz – 925 MHz	FDD
n101	1900 MHz – 1910 MHz	1900 MHz – 1910 MHz	TDD
n102	5925 MHz – 6425 MHz	5925 MHz – 6425 MHz	TDD
n104	6425 MHz – 7125 MHz	6425 MHz – 7125 MHz	TDD
n105	663 MHz – 703 MHz	612 MHz – 652 MHz	FDD

Table 5.2 NR operating bands in FR2.

Operating band	Uplink (UL) operating band BS receive UE transmit F_{UL_low} – F_{UL_high}			Downlink (DL) operating band BS transmit UE receive F_{DL_low} – F_{DL_high}			Duplex mode
n257	26,500 MHz	–	29,500 MHz	26 500 MHz	–	29 500 MHz	TDD
n258	24,250 MHz	–	27,500 MHz	24 250 MHz	–	27 500 MHz	TDD
n259	39,500 MHz	–	43,500 MHz	39 500 MHz	–	43 500 MHz	TDD
n260	37,000 MHz	–	40,000 MHz	37 000 MHz	–	40 000 MHz	TDD
n261	27,500 MHz	–	28,350 MHz	27 500 MHz	–	28 350 MHz	TDD
n262	47,200 MHz	–	48,200 MHz	47 200 MHz	–	48 200 MHz	TDD
n263	57,000 MHz	–	71,000 MHz	57 000 MHz	–	71 000 MHz	TDD

reception, enabling simultaneous communication in both directions. Others use time division duplex (TDD), where the same frequency is used for both uplink and downlink but at different times. Additionally, certain bands are designated for one-way communication, such as supplementary uplink (SUL) or supplementary downlink (SDL), depending on the requirements of the network.

The choice of operating frequency band is closely tied to the specific application scenario. In general, low-frequency bands (below 1 GHz) offer broader coverage and can penetrate obstacles such as building walls more effectively than higher-frequency bands. This makes them ideal for long-range communication and rural or suburban deployment. On the other hand, high-frequency bands, such as those in the mmWave spectrum (above 20 GHz), provide significantly higher bandwidth but are more prone to signal attenuation over long distances and have difficulty penetrating physical barriers. As a result, high-frequency bands are best suited for high-bandwidth, short-range applications, like dense urban areas or indoor environments.

Furthermore, 5G New Radio (NR) supports carrier aggregation, a technology that combines multiple frequency bands to create a wider effective bandwidth. This capability allows for simultaneous connections across different frequency bands, which is particularly beneficial for high-bandwidth applications such as augmented reality (AR) and virtual reality (VR). By using carrier aggregation, 5G networks can deliver enhanced performance and improved user experiences for data-intensive applications.

In addition to carrier aggregation, the use of supplementary carriers can further optimize network performance. By adding extra frequency bands to the primary carrier, the network can offload traffic when congestion occurs, thereby improving overall efficiency. For example, if the primary uplink or downlink carrier becomes congested, a 5G network can seamlessly switch to a supplementary carrier, ensuring continuous service and optimal utilization of available spectrum resources. This dynamic allocation of frequency bands contributes to the enhanced reliability and flexibility of 5G networks, especially in high-demand environments. For more detailed information on channel configuration, refer to Ref. [6].

5.3 Core Requirements

In line with the IMT-2020 framework, the three core requirements for 5G—Enhanced Mobile Broadband (eMBB), Massive Machine-Type Communications (mMTC), and Ultra-Reliable Low-Latency Communications (URLLC)—each address unique applications and use cases that leverage 5G's advanced capabilities:

(a) Enhanced Mobile Broadband (eMBB)
eMBB is primarily focused on delivering high data rates, improved connection densities, and enhanced user mobility. This allows for more immersive and data-intensive applications to function seamlessly on mobile networks.

4K/8K Video Streaming – With eMBB, mobile networks can deliver ultra-high-definition video streaming, such as 4K or 8K video, with minimal buffering and interruptions, even in crowded areas like sports stadiums or concerts.

Augmented and Virtual Reality (AR/VR) – eMBB enables the smooth delivery of AR and VR content to mobile users. For instance, a tourist could use AR glasses to receive real-time historical data superimposed on landmarks, while gamers could participate in VR environments that require fast, seamless data transmission.

Cloud Gaming – With the low latency and high data rates provided by eMBB, cloud-based gaming services such as Xbox Cloud Gaming or Google Stadia can deliver high-quality graphics and real-time multiplayer interactions to users without the need for powerful local hardware.

Technologies like massive MIMO (which increases spectral efficiency by using a large number of antennas) and beamforming (which directs signals precisely towards users) are critical in supporting these high-speed, data-intensive applications.

(b) Massive Machine-Type Communications (mMTC)
mMTC supports the connectivity of vast numbers of IoT devices, each transmitting small amounts of data. This makes it ideal for environments where a large number of devices must communicate autonomously, often without human intervention.

Smart Cities – In a smart city, thousands of IoT devices, such as traffic sensors, streetlights, and surveillance cameras, communicate with one another and central systems. mMTC enables the smooth operation of such infrastructure, improving traffic management, energy usage, and public safety.

Industrial Automation – Factories increasingly rely on IoT sensors and devices to monitor machines, optimize workflows, and reduce downtime. With mMTC, 5G networks can handle the enormous data traffic generated by these devices, ensuring real-time coordination of robotic arms, conveyor belts, and other automated systems in smart factories.

Agriculture and Precision Farming – In modern agriculture, sensors are deployed to monitor soil moisture, weather conditions, and crop health across large areas. mMTC allows these sensors to communicate with each other and send data to farmers in real-time, enabling better decision-making and increased productivity in precision farming.

Network slicing, a technology that partitions the network into virtualized segments optimized for specific use cases, is instrumental in ensuring that high-performance, low-power IoT devices function efficiently, without interference from other applications on the network.

(c) Ultra-Reliable Low-Latency Communications (URLLC)
URLLC is designed for applications that demand extremely low latency (often below 1 millisecond) and high reliability. This is crucial for safety-critical systems where even the slightest delay could have significant consequences.

Autonomous Vehicles – Self-driving cars rely on instant communication with other vehicles, traffic lights, and roadside infrastructure to navigate safely and avoid collisions. URLLC ensures that these communications occur with near-zero latency, enabling real-time decision-making and enhancing road safety.

Remote Surgery – In telemedicine, URLLC makes it possible for a surgeon to perform a procedure on a patient remotely, controlling robotic instruments in real-time. The ultra-low latency provided by URLLC ensures that the surgeon's movements are accurately mirrored by the robot, minimizing the risk of delay-related errors.

Industrial Robotics – In high-precision manufacturing environments, industrial robots must coordinate with extreme accuracy. URLLC ensures that commands from central systems are executed with near-instantaneous response times, allowing for real-time monitoring and control of robotic processes in environments like semiconductor manufacturing or assembly lines.

Edge computing, which processes data closer to the source (e.g., at the base station or on-site server) rather than relying on centralized cloud data centers, is crucial for reducing latency in URLLC use cases. This ensures that data doesn't need to travel long distances, minimizing delays and allowing critical systems to respond in real time.

These three requirements—eMBB, mMTC, and URLLC—not only define the technical capabilities of 5G but also open the door to transformative applications across various industries. Whether it's enabling immersive media experiences,

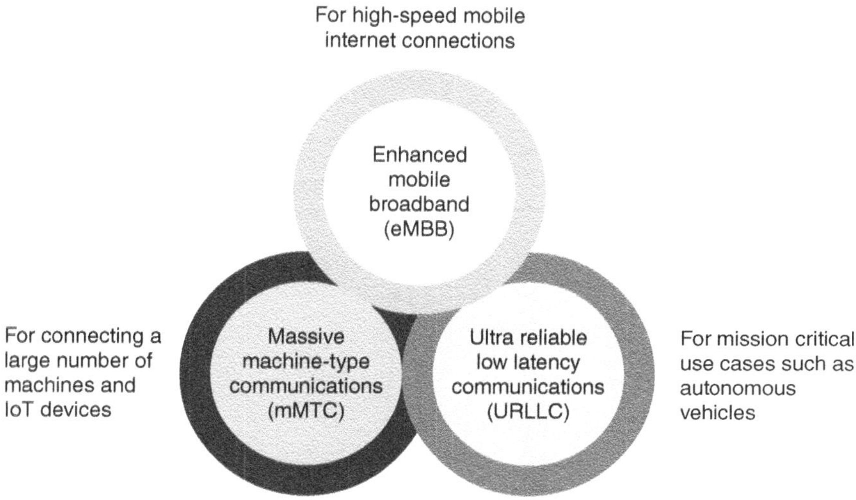

Figure 5.2 5G standard requirements.

managing massive IoT ecosystems, or ensuring reliable real-time communications in critical scenarios, 5G's versatility and performance improvements are key to realizing the next generation of connected technologies. Figure 5.2 illustrates how these three use cases form the foundation of 5G's diverse applications.

5.4 New Radio Access Technology

As discussed earlier, 3GPP introduced a new radio access technology in Release 15 known as 5G New Radio (NR), based on orthogonal frequency division multiplexing (OFDM). This technology is designed to support Enhanced Mobile Broadband (eMBB) and Massive Machine-Type Communications (mMTC) applications, enabling high-speed data transmission and low-latency communication. Succeeding 4G LTE, 5G NR brings several key improvements, such as massive MIMO and beamforming technologies. Below, we explore these key technologies in NR.

5.4.1 Orthogonal Frequency-Division Multiplexing

OFDM is a signal modulation technique that increases data rates and enhances spectral efficiency by dividing the main carrier into multiple subcarriers. Based on the frequency division multiplexing scheme, the subcarrier signals are arranged orthogonally to prevent interference. Figure 5.3 shows a multicarrier arrangement, illustrating that subcarriers are spaced apart without overlapping in the frequency domain.

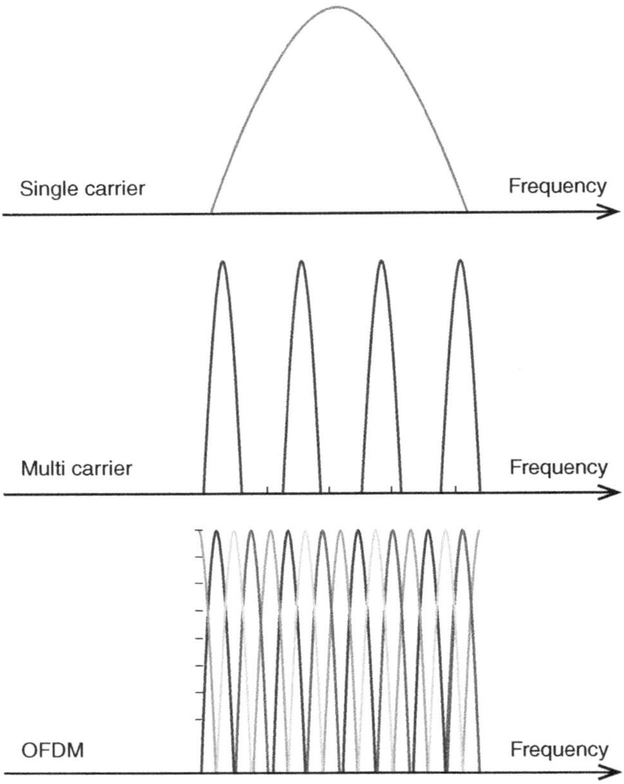

Figure 5.3 Multi-subcarrier signals arrangement in comparison to a single carrier and multicarrier arrangement. Source: Choudhary and Janyani [7]/with permission of Springer Nature.

In OFDM, at any one frequency, when a subcarrier is received at maximum strength, adjacent subcarriers carry zero-signal magnitude. This ensures that subcarriers do not interfere with each other, effectively eliminating crosstalk between signals.

Key advantages of OFDM include:

High Data Rates: By transmitting multiple subcarriers simultaneously in parallel, OFDM enables high data speeds. More subcarriers can be added as needed to support increased data transmission.

Multipath Fading Resistance: OFDM uses cyclic prefixes to minimize the effects of multipath fading, which can weaken signals and cause errors. Multipath fading occurs when signals are reflected by obstacles like buildings or trees, causing them to arrive at the receiver via different paths and at different times.

Low Latency: The parallel transmission of subcarriers and the use of cyclic prefixes to mitigate multipath fading help reduce latency, making OFDM suitable for real-time applications such as video streaming, virtual reality (VR), and augmented reality (AR).

5.4.2 Massive MIMO (Multiple-Input Multiple-Output)

MIMO involves the use of multiple antennas for transmitting and receiving wireless signals, which was also a feature in 4G LTE. A basic MIMO system operates in a 2T2R configuration, meaning two transmitters and two receivers are available. However, such a 2 × 2 MIMO setup offers relatively low capacity. Figure 5.4 illustrates both 2 × 2 and 4 × 4 MIMO configurations.

In 5G, massive MIMO refers to configurations with a large number of antenna elements, such as 16 × 16, 32 × 32, or even 64 × 64. Due to limitations in user equipment (UE), a 4 × 4 MIMO configuration is often used at the receiver end. In addition to higher data rates, massive MIMO increases capacity through techniques like multiuser MIMO, which allows multiple users to be served simultaneously, making it especially useful in high-density areas.

5.4.3 Beamforming

Beamforming is essential for supporting massive MIMO configurations. Multiple antenna elements are arranged in arrays, allowing them to transmit focused, narrow radio beams with higher directivity compared to traditional single-antenna setups. Beamforming enables the network to direct radio signals specifically toward individual UEs, improving signal strength and coverage.

Beamforming offers several advantages:

Improved Coverage: With more narrow beams available, massive MIMO systems can provide better coverage.

Reduced Interference: The focused nature of the beams helps avoid interference from signals transmitted by other antennas. Figure 5.5 shows a simplified diagram of antenna beam patterns from an antenna array.

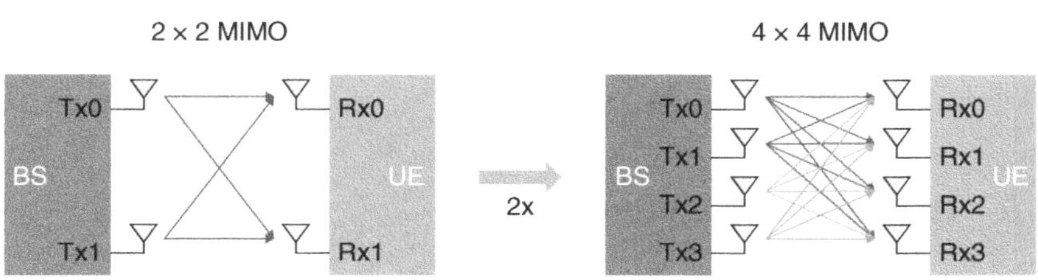

Figure 5.4 Example of MIMO configurations: 2 × 2 and 4 × 4. Source: From 5G Indonesia https://www.5g-indonesia.com/2020/08/penjelasan-lengkap-massive-mimo-5g.html.

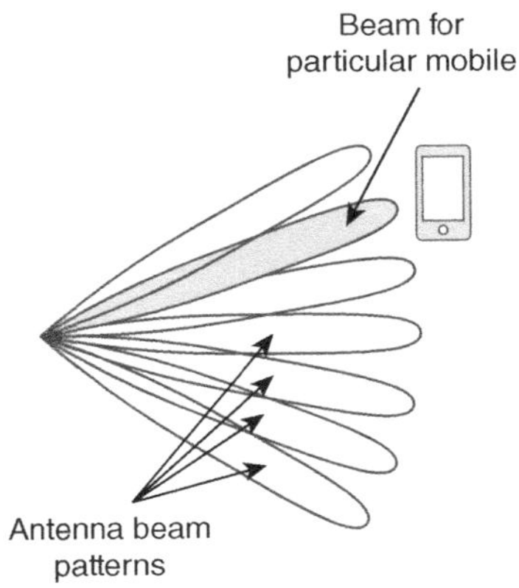

Figure 5.5 Antenna beam targeting particular user equipment. Source: From https://www.electronics-notes.com/articles/connectivity/5g-mobile-wireless-cellular/massive-mimo.php.

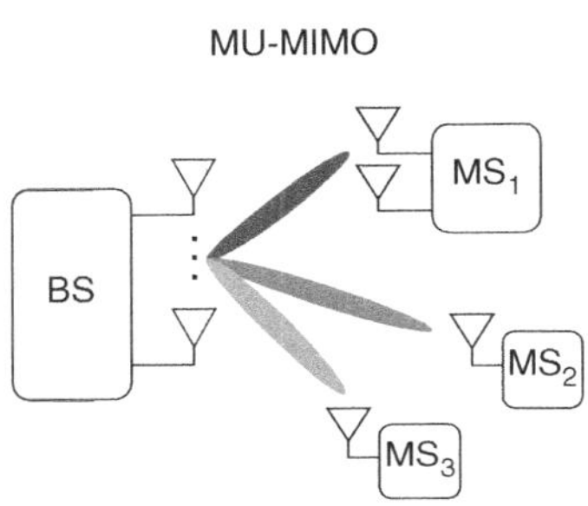

Figure 5.6 Multiuser MIMO configuration. Source: A. Shaikh and M. J. Kaur, 2019/with permission of IEEE. https://www.everythingrf.com/community/what-is-mu-mimo.

5.4.4 Multiuser MIMO

While earlier discussions focused on how a 5G base station can serve a single user, multiuser MIMO allows for the simultaneous servicing of multiple UEs located in different areas. This is achieved by pooling resources such as signal frequency or timing for all UEs in the group. Since UEs in different physical locations use different spatial paths, they do not interfere with one another, improving user capacity.

However, UEs positioned too close to each other should not be grouped, as this could result in interference. As shown in Figure 5.6, users like MS3 and MS1 can be served simultaneously because they are sufficiently separated, preventing interference.

5.5 Network Architecture

The architecture of a 5G network consists of two main components: the access network and the core network. A generic 5G system covers a radio access network

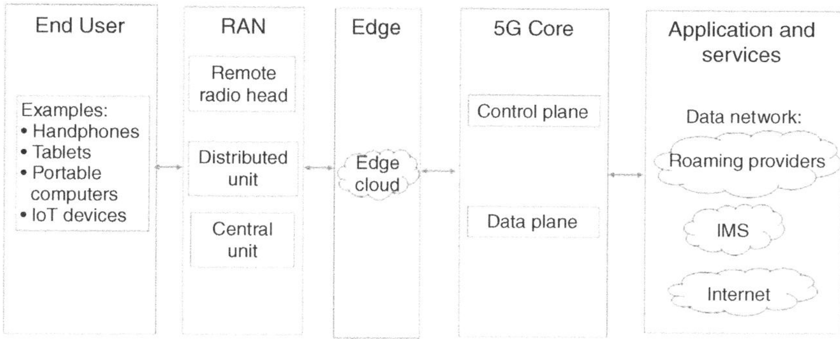

Figure 5.7 A generic 5G network.

(RAN), an edge, and a core, connecting user equipment (UE) to a data network, enabling various applications and services (Figure 5.7).

The key components of the 5G network are briefly outlined below, with more details on their software architecture provided in the next chapter.

5.5.1 Next Generation Radio Access Network

User devices, also known as UEs, connect to the RAN via a radio link. Traditionally, UEs were mobile phones or tablets, but today they include a wide range of devices like sensors, drones, robots, and other unmanned vehicles. The RAN may also connect to multiple access technologies, including LTE (E-UTRA).

NG-RAN comprises the network of 5G radio towers and cells, supporting both mobility and security. It plays a key role in managing radio spectrum resources to ensure efficient delivery of high-quality service to users. The RAN includes base stations known as next-generation Node B (gNB), which are divided into centralized units (CU) and distributed units (DU), connected via the F1 interface (see TS38.401 for more details). An NG-RAN can consist of:

gNB: Provides NR (5G) user plane and control plane protocol termination for UEs.

ng-eNB: Provides E-UTRA (4G) user plane and control plane protocol termination for UEs.

The overall NG-RAN structure is depicted in Figure 5.8. Connections between gNB and ng-eNB use the Xn interface, while 5G core network connections with gNB and ng-eNB are made through the NG interface.

These innovations in network architecture and radio technologies ensure 5G networks are equipped to deliver high-speed, low-latency, and high-capacity services for a wide range of applications.

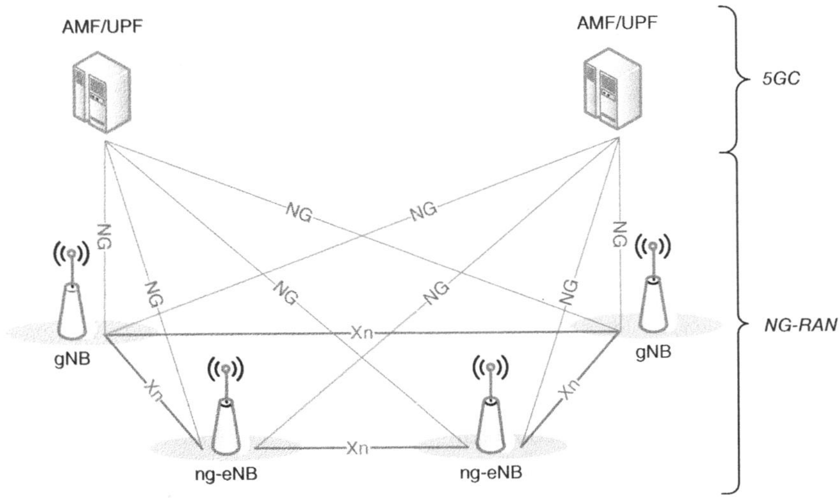

Figure 5.8 NG-RAN architecture. Source: [8]/https://www.etsi.org/deliver/etsi_ts/138300_138399/138300/15.08.00_60/ts_138300v150800p.pdf/3GPP.

5.5.2 5G Core

In this section, we will discuss the key features of a 5G core, which is divided into two functions: the control plane (CP) and the user plane (UP).

5.5.2.1 Control and User Plane Separation (CUPS)

The 5G core functions of CP and UP are separated and work independently. The CP handles the network signaling and management functions, while the UP handles data traffic. This arrangement allows flexibility and scalability in a 5G network, as CP and UP can be scaled up or down independently. In traditional non-5G networks, both CP and UP must be scaled up or down to meet the increasing or decreasing data traffic needs. In addition, CUPS can handle new services in a distributed and virtualized fashion by processing data around the edge of the 5G network, thus reducing latency. This is useful for applications such as AR, VR, and video streaming.

5.5.2.2 Service-Based Architecture (SBA)

Built upon a service-oriented architecture, each SBA service is arranged in a modular way and works as a standalone unit. Each service can be easily integrated with other services, making this architecture very useful for setting up new services. SBA architecture uses a set of service-based interfaces to connect different network functions. These interfaces are designed to handle voice, video, and data.

A key capability of SBA is to handle network slicing – a virtualization technique allowing different network services to work in a shared physical network infrastructure. For instance, the network can be divided into multiple virtual networks, each having its own characteristics and capabilities for handling use cases in eMBB, mMTC, and URLLC. By processing data closer to the source, we will improve the overall network performance. An SBA arrangement is shown below.

The key functions of the SBA are:

- User plane function (UPF)
- Access and mobility management function (AMF)
- Session management function (SMF)
- Authentication server function (AUSF)
- Network slice selection function (NSSF)
- Policy control function (PCF)
- NF repository function (NRF)
- Network exposure function (NEF)
- Unified data management (UDM)
- Application function (AF)

In addition to SBA, another 5G architecture is called reference point architecture. Figure 5.9 also shows dedicated interfaces between different functions:

- **N1:** Between 5G core and user equipment
- **N2:** Between 5G core and base station (part of RAN)
- **N3:** Between RAN and user plane function
- **N4:** Between session management function of the control plane and the user plane
- **N6:** Between user plane function and data network such as Internet
- **N9:** Between two user plane functions (i.e., intermediate I-UPF and the UPF session anchor)

More information about SBA and the interfaces are discussed in the next chapter.

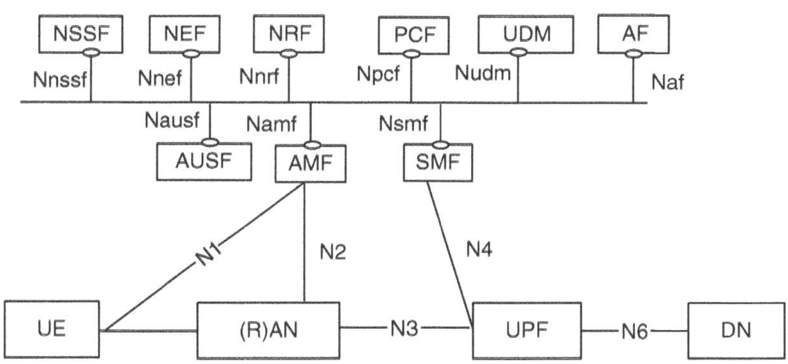

Figure 5.9 5G system software-based architecture [9]/https://www.etsi.org/deliver/etsi_ts/123500_123599/123501/17.05.00_60/ts_123501v170500p.pdf/3GPP.

5.6 Performance Improvement

The emerging 5G/6G AI and metaverse applications are in dire need of innovative architectures and microarchitectures delivering hundreds of Gb/s bandwidth and <1 ms latency in the era of computing and network convergence (CNC). Building a CNC-centric computing architecture demands a profound paradigm shift, incorporating network devices as a vital part of the computing engine and not merely as external peripheral devices. The key design focus is to ensure the most efficient architecture that can optimize data flows and eliminate any unnecessary data movement in computing and network software/hardware infrastructures.

In this section, we examine the characteristics of a wide range of demanding computing and network tasks in a number of integrated scenarios. We present a novel distributed computing architecture – called Smart_xPU[1] – that optimizes the data paths among all data producers and consumers. Smart_xPU can reduce power consumption by 66.7% and 50.0% compared to the two traditional architectures, respectively. In addition, Smart_xPU is fully compatible with the de facto OpenCL standard, and its efficient data flows can markedly support better bandwidths and latencies. Our research team aims to further optimize the Smart_xPU architecture in the near future.

5.6.1 Computing and Network Convergence

CNC has recently received unprecedented attention from academia and industry alike, partly due to the importance of many emerging 5G AI applications in which a large amount of data from a network must be processed in a timely fashion [10]. CNC has become a research focus in many prominent conferences in 2022, including the World Artificial Intelligence Conference, the World of Computing Conference, and the International Conference on Service Science.

The China Mobile research group, led by Wang Xiaoyun, observed that the network functions are gradually migrating from providing communication services to providing information and data processing channels. As such, it is meaningful to incorporate high-performance computing (HPC) elements within the network architecture (https://baijiahao.baidu.com/s?id=1742909693943954278&wfr=spider&for=pc). Meanwhile, the Chinese government has fully committed to its “double carbon” goals, reaching peak carbon use by 2030 and becoming carbon neutral by 2060 [11]. Accordingly, future CNC architecture is required to achieve exceptionally high performance with incredible power efficiency.

Some of the world’s largest data centers are already consuming up to 100 MW, but the ever-increasing 5G cloud applications would need additional power

1 Smart_xPU is a registered trademark of Inspur Group.

dissipation. It is paramount to minimize and save every bit of power in every single entity. A power-efficient CNC architecture not only saves huge electricity costs but can also help China achieve its "double carbon" objectives, providing meaningful and positive social impacts (https://news.sina.com.cn/o/2021-05-11/doc-ikmyaawc4726524.shtml). Today, most researchers believe an efficient CNC architecture would require seamless integration of cloud, edge, network, and devices, which in turn would require ongoing and broad global collaborative research efforts.

The computing performance is often inseparable from the network performance in the CNC era. For instance, many deep neural network (DNN) inferencing projects may require complicated computation with high throughput and low latencies, but the performance bottleneck can be located within the computational units, the network, or both. Consequently, there is an urgent need for design architects to come up with new DNN computing architectures aiming at achieving low latency and high throughput with flexibility [12–14].

One such solution came from Microsoft. Microsoft unveiled its Brainwave project at Hot Chips 2017 [15, 16], an acceleration architecture for ultra-large-scale data centers. Their principal idea is to integrate a high-performance reconfigurable DNN accelerator based on field programmable gate array (FPGA) with a central processing unit (CPU) and network card through PCIe. This is to realize low-delay and high-bandwidth communication between any nodes in the network. The number of FPGAs in a single node can be expanded as needed to avoid residual capacity, improving the overall utilization of the data center to a certain extent. Nevertheless, this architecture has an inherent weakness. Because the architecture is CPU-centric, this means that CPU must be needed within each node, and there is a tight coupling among CPU, FPGA, and network components. This can reduce design flexibility and impose unnecessary constraints on the deployment and scheduling of various computing resources.

As the world's leading and largest provider of AI computing platforms, Inspur focuses on four key design objectives: performance, algorithms, power consumption, and AI computing eco-system (https://http://3g.163.com/dy/article/H2P66L0205118A91.html). The overall design involves many technical disciplines, such as materials, thermodynamics, battery technology, fluid mechanics, and chemistry. As an illustration, an Inspur AI server may encompass 10,000+ components, including 50+ customized chip designs, 30+ technical initiatives, and 100+ communication protocols. In addition, the production involves more than 30 overall procedures, 150 manufacturing and handling processes, as well as 280 quality assurance and compliance processes. It is indeed a complex project to ensure the quality of each Inspur AI platform system (https://m.gmw.cn/baijia/2021-11/01/35278679.html).

The critical research considerations involve integrating complex computing architectures with various network topologies, supporting scale-up and scale-down systems, limiting dark silicon effects within each chip, optimizing performance while minimizing power consumption. To this end, a lightweight and fast reconfigurable computing engine and verification platform would be most interesting.

In this section, we will discuss an FPGA-based prototype to explore various design options in non-Von Neumann parallel computing architecture and present a novel distributed computing architecture known as Smart_xPU. Our main discussions include:

1) We present a CNC-centric computing architecture, eliminating the constraints within traditional Von Neumann IO architectures by combining control and data near network interfaces so that data are immediately processed without incurring unnecessary delays. Compared with two typical architectures, the number of data movements may be reduced by 83% and 50%, respectively, and the power consumption can be reduced by 66.7% and 50.0%, respectively.
2) Although Smart_xPU is decoupled from the CPU, it is fully compatible with the OpenCL programming standard, allowing an easy transition from transitional designs to the new architectural platform.
3) We implement an FPGA-based prototype to demonstrate and verify our design.

5.6.2 Related Works

Besides the Microsoft's Brainwave project described in the last section, there are also some related CNC architectures. For example, Nvidia has proposed four different GPUDirect technologies over the past decade, including GPUDirect shared memory [17], GPUDirect P2P [18], GPUDirect RDMA [19], and GPUDirect Async [20]. The comparison of these four architectures is shown in Figure 5.10.

Among these four architectures, GPUDirect Shared Memory was its first release. In this architecture, the source data are first copied from the graphic processing unit (GPU) memory to the local host memory. The CPU then instructs the local NIC to copy the data from the local host memory to a remote NIC. GPUDirect shared memory requires extra data movements, thus causing long latencies. Compared to GPUDirect shared memory, GPUDirect P2P is a major improvement as it can support zero-copy data transfers. GPUDirect P2P still requires the CPU's involvement, and it can only support data transfers via PCIe bus between GPUs within the same node. GPUDirect RDMA can support data transfers between GPUs in two different nodes, using the CPU to trigger the data transfers from GPU memory to the NIC via the PCIe bus. The previous three GPUDirect architectures all require CPU in data transfers, but GPUDirect Async

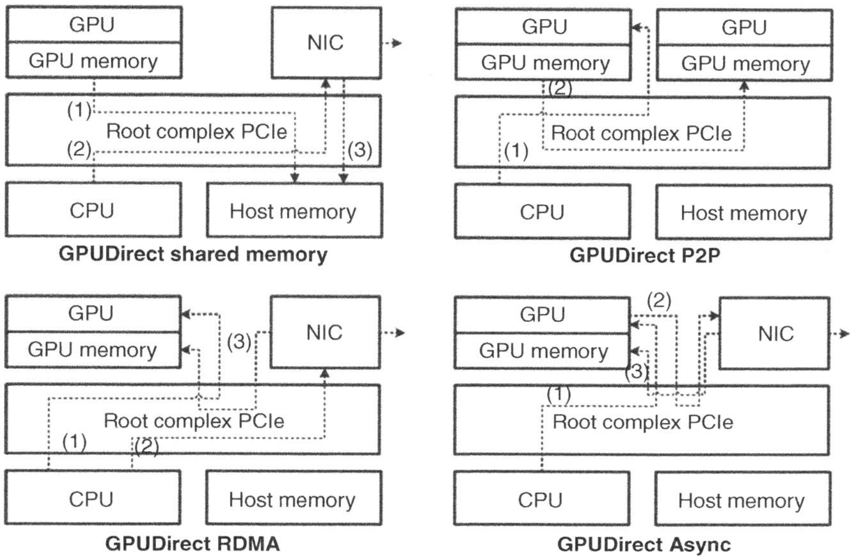

Figure 5.10 Comparison of data communication flows among four different Nvidia GPUDirect architectures: GPUDirect shared memory, GPUDirect P2P, GPUDirect RDMA, and GPUDirect Async.

uses GPU to trigger the NIC and does not involve CPU. Unfortunately, all the GPUDirect architectures have additional overheads due to redundant cross-chip transmission.

Based on Nvidia's software and hardware models, Oden proposed a different communication model known as the Global GPU Address Space (GGAS) [21]. GGAS achieves zero-copy data transfers among GPUs without the CPU's involvement on a custom RDMA-enable FPGA network interface card (NIC).

However, the CPU and GPU share the same NIC in a node, and the performance is bounded by the PCIe link. Besides, the actual bandwidth is less than one-third of the theoretical bandwidth due to the poor support of the end-to-end communication by the PCIe system on common chipsets. Furthermore, they conclude that allowing GPUs to autonomously send and receive network traffic is less efficient due to the poor performance of the single thread in the GPU and the PCIe overhead.

Daoud proposed a GPU-side RDMA library, GPUrdma [22], which implements high-throughput and low-latency communications for GPU kernels. However, special PCIe switches need to be used to avoid the significant difference between read and write speeds. It also requires huge efforts to modify the corresponding drivers for the special GPU and NIC due to the hardware constraints in the NVIDIA GPUs.

Silberstein proposed a networking abstraction for GPU kernel programs named GPUnet [23]. It provides socket-like high-level in-kernel APIs to bypass the CPU. The limitation of Nvidia GPUs makes a complete CPU bypass unpractical. In fact, it demands the assistance of the CPU to handle the network packages, such as the synchronization of flow control between the GPU and CPU.

It is obvious that all the above solutions have similar drawbacks that redundant cross-chip transfers exist in the data path. IBM proposed a network-based FPGA cluster [24]. By directly connecting FPGA to the data center network as an independent resource, it changes the traditional CPU-FPGA interface paradigm, decoupling the FPGA from the CPU, and releasing it from the CPU and PCIe bus. Unfortunately, IBM's proposal requires an overly complicated initialization process and can only support private FPGA programming logic APIs, which are incompatible with the existing programming paradigm.

As FPGA is an important heterogeneous computing component, FPGA manufacturers spare no effort to put forward a series of communication solutions [25, 26]. Inspur Group and Intel Corporation jointly demonstrated the Inter-Kernel Links (IKLs) solution [27], used for distributed direct communication between FPGAs, as shown in Figures 5.11 and 5.12. IKL enables communication between FPGAs without the involvement of PCIe and CPU, and the NIC only belongs to the accelerator function unit (AFU) in one FPGA device [28]. However, the implementation of this credit-based flow control and channel-based P2P communication model may drastically affect system scalability and interoperability. The IKL architecture has several important issues. First, IKL only focuses on data transmission

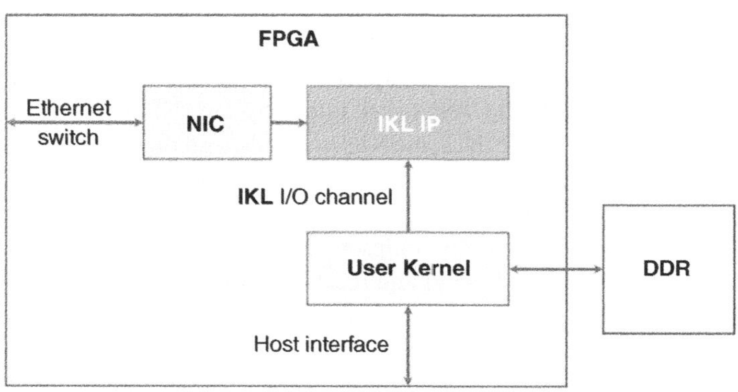

Figure 5.11 The key components within the Inter-Kernel Links (IKL) architecture, jointly developed by Inspur Group and Intel Corporation.

Figure 5.12 The distributed FPGA-based acceleration platform, jointly developed by Inspur Group and Intel Corporation, demonstrated in the Intel FPGA Technology Day (IFTD) 2019.

among FPGAs, and the communication path must be fixed in advance. Second, the number of communication links is limited by the number of ports, and the circuit frequency is determined by the actual User Kernel circuit. Finally, the data transmission between CPU and FPGA is ignored, creating a cooperation barrier for the different computing engines.

Meanwhile, Intel has revealed its XPU architectural vision, becoming the only chip vendor providing all four major controller chip types: CPU, GPU, FPGA, and system-on-chip (SOC). It also announced plans to launch "Falcon Shores XPU" – the processor architecture of fusion processors – in 2024 (https://community.intel.com/t5/Blogs/Products-and-Solutions/HPC/Delivering-a-Multi-Year-HPC-Roadmap/post/1395728?wapkw=Falcon%20Shores%20XPU; https://www.intel.com/content/www/us/en/newsroom/opinion/accelerated-innovations-sustainable-open-hpc.html#gs.nwdqqu). The "Falcon Shores XPU" family chips integrate Intel's x86 CPU and Xe GPU on the same Xeon chip and are expected to deliver more than five times improvements in performance per watt, computing density, memory capacity, and performance throughput.

The three major Chinese telecommunication companies are all actively promoting CNC research. In the 2022 Computing Network Integration Technology Seminar, the China Mobile Research Institute, led by Sun Tao, proposed a new CNC architecture [29], transferring part of computing tasks from the host side to the network side. This means that network devices such as switches, routers, smart NICs, and data processing unit (DPU) processing cards can use accelerated computation to improve the network throughput and reduce its latency and power consumption.

Kan Hongwei proposed a new research direction (https://dl.ccf.org.cn/video/videoDetail.html?_ack=1&id=6092696873895936; https://conf.ccf.org.cn/web/api/m9644563065535242241649985902214.action) by deploying various computing engines on the network/storage side and utilizing them as basic I/O devices within the architecture. In this architecture, some special system load can be diverted to eliminate the traditional CPU-centric mode, making the computing channels more efficient and eliminating many blocked transactions. In this approach, the deployment of distributed XPU chips used in a distributed server pool can effectively reduce the dependence on specific high-end chip processes and can balance the design trade-offs among performance, power consumption, flexibility, and efficiency.

5.6.3 Smart_xPU Design Methodology

Moore's law has been broken for more than a decade when the Drennan scaling was no longer valid below 20 nm silicon process technology. Beyond that point, each new technology process node can only provide marginal performance gain and power reduction. In the current post-Moore period, HPC platforms have to increasingly rely on novel architecture and complicated microarchitecture in order to mitigate the effects of interconnect and synchronization latencies, manifested in the memory wall, I/O wall, communication wall, and synchronization dispatch wall. These four walls are the major impediments to many emerging 5G/6G AI applications, which demand <1 ms latencies and 100s of Gb/s throughput. At the same time, there is a dire need to minimize computing and network power consumption as China is fully committed to its "double carbon" goals.

To this end, we have proposed a novel heterogeneous architecture, called Smart_xPU to address these latency, throughput, and power issues. The Smart_xPU platform represents a new system design methodology, supporting different types of chips – CPU, GPU, FPGA, SOC – from many different suppliers.

In the CNC era, the Smart_xPU architecture is tasked to provide better PPA (performance, power, area) metrics in the 5G/6G AI applications compared to traditional architectures with the same process technology. Specifically, the Smart_xPU design methodology encompasses four key design optimization flows (Figure 5.13). In this section, we will discuss these design optimization flows and analyze our Smart_xPU.

5.6.3.1 Data Flow Optimization

In the CNC era, data transmission – not data processing – plays a predominant role in determining the power consumption in a typical computing platform. In a typical 7 nm process SOC, data transmission may contribute around 64% of the overall chip power dissipation. There are three main interconnect types – on-board

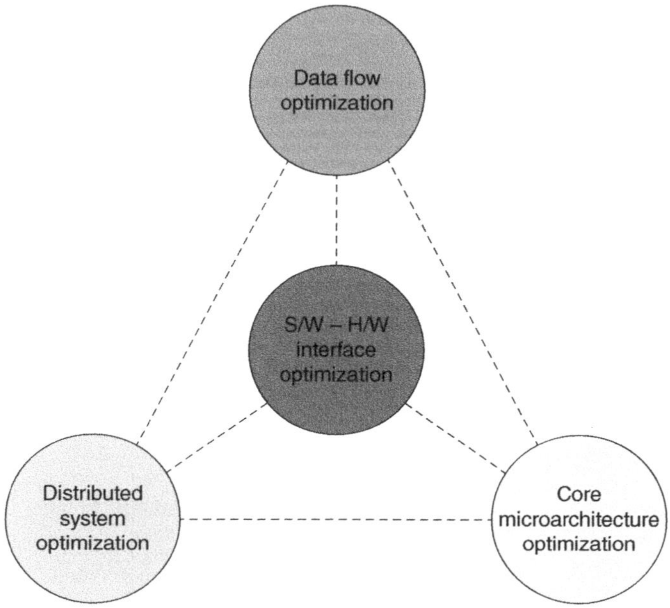

Figure 5.13 The Smart_xPU architectural design follows a novel design methodology, encompassing four different design optimization flows: data flow optimization, distributed system optimization, core microarchitecture optimization, and software/hardware interface optimization.

interconnect, on-package interconnect, and on-die interconnect – in a computing device. Intel's research team has estimated the per-bit energy consumption in data transmission via these three interconnect types, as shown in Table 5.3 [30]. Among these three types, on-die interconnect is usually the most power efficient, and on-board interconnect is usually the least power efficient.

Table 5.3 Comparison of per-bit energy dissipation in data transmission via on-board interconnect, on-package interconnect, and on-die interconnect.

	On-board interconnect	On-package interconnect	On-die interconnect
Length (m)	<1	~0.001	~0.0001
Energy (pJ/bit)	20	<1	0.1
Bus standards	PCIe, DDR, etc.	UIB, AIB, etc.	IOSF, AMBA, etc.

In general, on-die interconnect is more efficient and faster than on-package interconnect, and on-package interconnect is more efficient and faster than on-board interconnect.

Regardless of some subtle differences among existing architectures, we can use mainstream general-purpose servers and AI-specific servers as our illustrations. As shown in Figure 5.14, the on-chip and off-chip data transfers between two GPUs are discussed in three typical scenarios:

- Figure 5.14a shows a typical server with a GPU accelerator. With CPU assistance, the GPU copies data from GPU memory (GPU Mem) to system memory (System Mem) via the PCIe bus. To allow network transmission, a context switch is required, and the data is copied once in System Mem. Finally, the CPU sends the data from System Mem to the NIC via the PCIe bus. The data transfers in the recipient node are similar but in reverse order.
- Figure 5.14b shows a dedicated GPU with a dedicated network adapter. Using the Nvidia GPU Direct Async system, the actual data flow transmission process

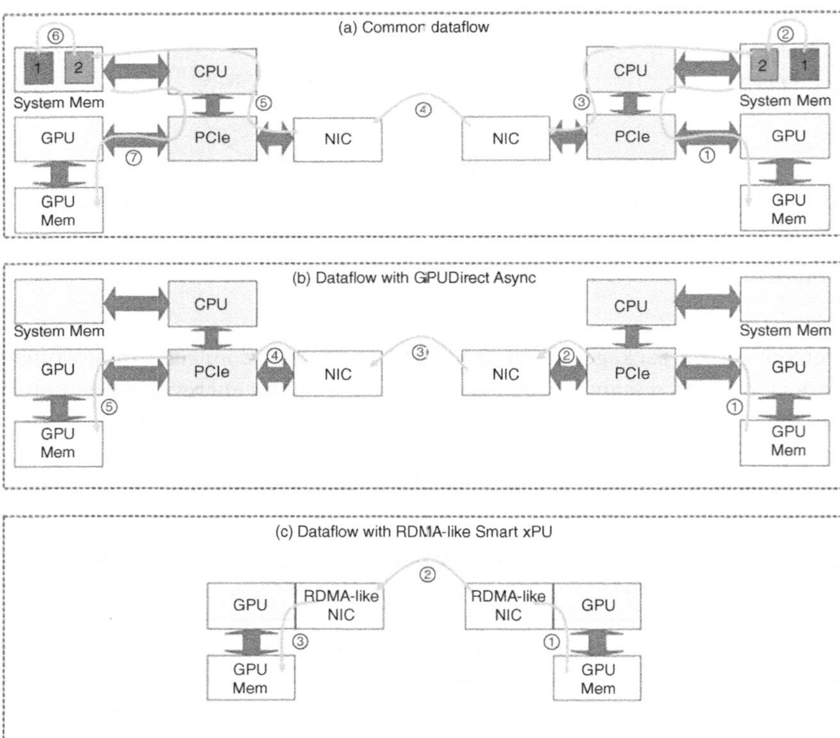

Figure 5.14 The on-chip and off-chip data transfers between two GPU nodes are discussed in three typical scenarios: (a) a typical server with a GPU accelerator, (b) a dedicated GPU with a dedicated network adapter, and (c) a CNC computing architecture with optimized data flow.

does not require CPU involvement. The GPU directly notifies the NIC, and the NIC reads the data from GPU Mem via the PCIe bus.

- Figure 5.14c shows a CNC computing architecture with optimized data flow. It treats the network as an essential I/O device in a system, supporting a direct GPU data communication channel, and eliminating two PCIe accesses and two DDR read–write accesses with tight coupling of the computing engine and network device.

The power consumptions (per data bit) for external data read/write operations of GPU Mem and System Mem are defined as P_{gpuMem} and P_{sysMem}, respectively. The power consumption (per data bit) of data flow via the PCIe bus is defined as P_{pci}. The power consumption (per data bit) of data flow via the network is defined as P_{net}. The power consumption (per data bit) of DDR access is defined as P_{inDDR}.

In all distributed system architectures, the power consumption (per data bit) to send data from one network node to another, P_{net}, is essential. However, this value depends on the network topology and varies greatly from one system to another. For simplicity, this section does not consider cross-node network communication.

Accordingly, we get:

$$\begin{cases} P_i = P_{OnBoard}, i \in \{gpuMem.sysMem, pci\} \\ P_j = P_{OnDie}, j \in \{inDDR\} \end{cases} \tag{5.1}$$

Thus, the unit power consumption of data flow in the three cases shown in Figure 5.14 is estimated as:

$$\begin{cases} P_{(a)} = 2 * P_{gpuMem} + 2 * P_{sysMem} + 2 * P_{pci} + 2 * P_{inDDR} \\ P_{(b)} = 2 * P_{gpuMem} + 2 * P_{pci} \\ P_{(c)} = 2 * P_{gpuMem} \end{cases} \tag{5.2}$$

Substituting typical values in Eq. (5.2), we get:

$$\begin{cases} P_{(a)} = 6 * P_{OnBoard} + 2 * P_{onDie} = 120.2\text{pJ per data bit} \\ P_{(b)} = 4 * P_{OnBoard} = 80\text{ pJ per data bit} \\ P_{(c)} = 2 * P_{OnBoard} = 40\text{ pJ per data bit} \end{cases} \tag{5.3}$$

Compared with traditional architecture and GPU Direct Async architecture, a CNC architecture with optimized data flow reduces the theoretical number of data movements by 75% and 50%, respectively. Similarly, the theoretical power consumption can also be reduced by 66.7% and 50.0%, respectively. This is very significant in terms of power and latency reduction. As such, our distributed Smart_xPU architecture is based on the CNC architecture as shown in Figure 5.14c.

5.6.3.2 Distributed System Optimization

The overall performance of a parallel distributed system depends on many factors, including the type of applications, performance of each node, power budget at each node, network latencies, and network topology. When designing and producing computing equipment, traditional manufacturers usually prioritize the performance of computing equipment in the beginning, assuming some theoretical performance based on each computing equipment. Unfortunately, the network can drastically slow down the whole system's performance and increase the system's power consumption. Accordingly, manufacturers would then add equipment to improve the weak communication link. For example, a smart network interface card (SmartNIC) [31] improves network performance and supports software-defined network (SDN), while a DPU [32] offloads many computing tasks, such as storage, security, and hypervisor.

Considering the complexity and imbalance of current network infrastructure, network communication based on CNC needs to be compatible with widely used communication means, such as high-speed Ethernet, RDMA-like Ethernet, and 5G/6G. Classical RDMA technologies based on IB protocol have the trend of being compatible with high-speed Ethernet [33, 34]. For example, RoCEV2 (RDMA over Converged Ethernet) implements an RDMA architecture based on UDP/IP protocol, and iWARP (Internet Wide Area RDMA Protocol) implements an RDMA architecture based on TCP/IP protocol.

Although these three protocols are all standardized, special network equipment is needed to run them because they significantly differ from the traditional network protocol stack. In addition, 5G/6G wireless networks already have their own standards and ecosystems. The integration of computing and network equipment should take compatibility and existing infrastructure into consideration.

Consider the Smart_xPU architecture. On the one hand, the inherent advantages of CNC can be fully utilized by reducing communication delay, improving transmission throughput, accelerating communication protocol processing, and so on. On the other hand, the passthrough between the computer and the network can be realized by eliminating the extra data flow overhead caused by route detouring and protocol conversion. Adopting the general principle of compatibility over innovation, maximizing the use of existing network infrastructure resources – a major CNC goal – is to design a proper interface module of CNC computing architecture. Accordingly, the following problems should be considered to realize the interface enabled with the integration of computing and network:

- Reposition its network function based on the new computing architecture; i.e., the essence of communication demand is the addressing access of data between engines;
- The design of CNC-based Smart_xPU needs to consider the strategy of adapting a national-wide network. The original link layer of the network needs to be retained, and the network layer needs to be compatible with the existing mainstream network infrastructure so as to be modified and enhanced to serve the

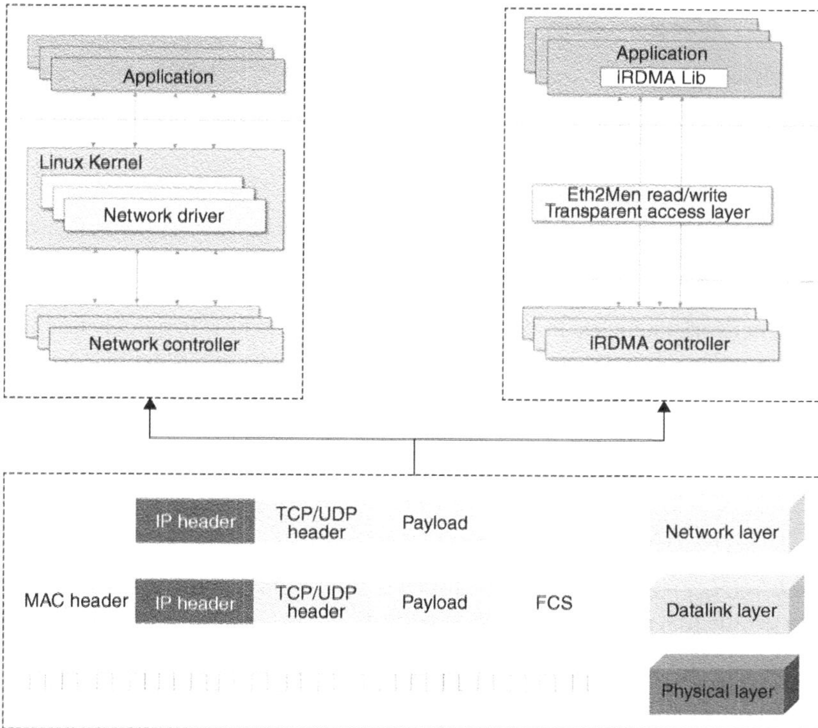

Figure 5.15 The data path of traditional Ethernet and direct addressing.

computing engine based on CNC and decouple the computing tasks on the traditional transmission layer (Figure 5.15).

- In this section, the Smart_xPU architecture integrates the computing engine and communication network module into a single chip. Using the "xPU over Ethernet" module with RDMA-like direct addressing and common network stack, Smart_xPU is compatible with traditional Ethernet-enabled devices. It can also realize direct input to the computing engine using RDMA, eliminating redundant switching chips in the communication path and removing extra PCIe accesses in a traditional distributed heterogeneous system. This points to an architecture that fully decouples CPU from xPU in a distributed computing system. As shown in Figure 5.16, computing, network, and data access functions are all tightly coupled in the xPU architecture.

When the number of the xPU nodes increases beyond a certain number, the overall performance will hit a diminishing return because the data communication costs can increase exponentially. The optimum number of nodes depends on

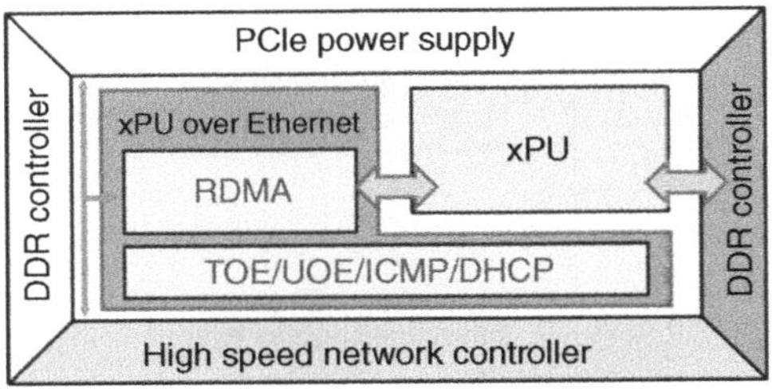

Figure 5.16 The xPU architecture with tightly coupled communication and computing units.

many factors, such as cost per node, target performance per node, network topology, and network latencies. In this subsection, we discuss how to quickly find the optimal number of nodes using a cost-performance objective function (N) and a few samples of performance measurement. The optimum number of N occurs when function $^{(1)}(N)$ is near 0. However, N is a discrete number, so the first derivative cannot be calculated directly. Fortunately, the first derivative can be calculated using Taylor's expansion. Using Taylor expansion, we get:

$$f(N_{i+1}) = f(N_i) + f^{(1)}(N_i)h + \frac{1}{2!}f^{(2)}(N_i)h^2 + O(h^2)$$

where $h = (N_{i+1} - N_i)$ is the sample distance. By simplifying the above formula, we can get the first derivative of $f(N)$ at point x_i:

$$f^{(1)}(N_i) = \frac{1}{h}(f(N_{i+1}) - f(N_i)) + O(h)$$

The derivative calculated from the above equation may not be accurate enough. To improve accuracy, the derivative of the above equation can be calculated again, and then:

$$f^{(2)}(N_i) = \frac{1}{h^2}(f(N_{i+2}) - 2f(N_{i+1}) + f(N_i)) + O(h)$$

By bringing this expression into the original expansion, we can make the first derivative more accurate.

$$f^{(1)}(N_i) = -\frac{1}{2h}[f(N_{i+2}) - 4f(N_{i+1}) + 2f(N_i)] + O(h^2)$$

If needed, the accuracy can be further improved by repeating the same Taylor expansion.

5.6.3.3 Core Microarchitecture Optimization

In this section, the xPU cores are implemented as RISC-V GPGPU in our Smart_xPU architecture. Each xPU core extends the RISC-V instruction set with a modified microarchitecture to support GPU SIMD capacity. Besides SIMD instructions, the xPU core also supports standard RISC-V instructions to run programs that are inherently sequential, like communication, task scheduling, and data compression.

In practice, the xPU cores need to handle a diversity of tasks; some of them can be easily parallelized, while others cannot be parallelized. It is, therefore, difficult to achieve satisfactory results after using common technologies to improve processor performance, such as multithreading and superscalar [35]. Inspired by some mature asymmetric multicore processor systems using large and small cores, we adopt a similar approach with our RISC-V GPGPU cores. Amdahl's law is used to analyze the optimal microarchitecture to maximize the efficiency of our computing engine. The general description is as follows. The common RISC-V-based GPGPUs are basically based on a symmetric multicore processor system, i.e., each processor core has the same size and performance. In this case, Amdahl's law can be expressed as:

$$S(p,n) = \frac{1}{(1-p) + \frac{p}{n}}$$

where p represents the portion in the task that can be parallelized ($p < 1$), n represents the number of xPU cores, and each core comprises some basic hardware unit. It is apparent that $\lim_{n\to\infty} S(p,n) = \frac{1}{(1-p)}$. In other words, the overall performance is always limited by the sequential portion of the task.

Assuming a fixed die size and process technology, the sequential and parallel parts of the task can be identified and modularized. Each basic hardware unit has a performance of 1, and a single chip of fixed size and process can hold a total of N basic hardware units. The theoretical performance of a core is basically positively correlated with the number of basic hardware units that make up the core. In a symmetric multicore processor, each core consists of w basic hardware unit cores, whose performance is $perf_{sym} = (w)$.

In an asymmetric architecture, the number of large cores is fixed at 1 because the execution time of a sequential task does not improve when the number of cores increases. Suppose there are u basic hardware units used to form the single large core and v basic hardware units used to form a small core, with a total of m small cores. Therefore, the performance of large and small cores can be expressed as:

$$\begin{cases} perf_{big} = perf\,(u) \\ perf_{small} = perf\,(v) = perf\left(\frac{N-u}{m}\right) \\ u + m \times v = N \; m, u, v \in Z^{+} \end{cases}$$

Using the inverse Amdahl's law, i.e., the higher the acceleration rate, the less time it takes to complete the task, the time required to complete the task under symmetric and asymmetric systems are:

$$\begin{cases} T_{sym} = \frac{1-p}{perf\,(w)} + \frac{p}{perf\,(w) \times \frac{N}{w}} \\ T_{asym} = \frac{1-p}{perf\,(u)} + \frac{p}{perf\,(u) + perf\left(\frac{N-u}{m}\right) \times m} \end{cases}$$

According to S. Borkar [36], $(k) = \sqrt{k}$, i.e., k basic hardware units can obtain the serial performance improvement of $\sqrt{k}$, so the time for symmetric and asymmetric architecture conditions is:

$$\begin{cases} T_{sym} = \frac{1-p}{\sqrt{w}} + \frac{p}{\frac{N}{\sqrt{w}}} \\ T_{asym} = \frac{1-p}{\sqrt{u}} + \frac{p}{\sqrt{u} + \sqrt{\frac{N-u}{m}} \times m} \\ \Delta T = T_{sym} - T_{asym} = (1-p) \times \left(\frac{1}{\sqrt{w}} - \frac{1}{\sqrt{u}} \right) + p \times \left(\frac{\sqrt{w}}{N} - \frac{1}{\sqrt{u} + \sqrt{N-u} \times \sqrt{m}} \right) \end{cases}$$

Taking into account the overhead of transforming the architecture, a single big core is employed in the asymmetric architecture, and the rest of the chip is filled with small cores. Therefore $=u$, we can get:

$$\Delta T = p \times \left(\frac{\sqrt{w}}{N} - \frac{1}{\sqrt{w} + \sqrt{N-w} \times \sqrt{m}} \right)$$

Considering that small cores must have lower performance than the big cores in the asymmetric architecture, we can get $\frac{N-u}{m} < u, m \geq \frac{N}{w'}$ so:

$$\frac{N}{\sqrt{w}} < \sqrt{w} + \sqrt{(N-w) \times m}$$

That is,

$$\frac{\sqrt{w}}{N} - \frac{1}{\sqrt{w} + \sqrt{(N-w) \times m}} > 0$$

The above results show that the accelerating ratio of the asymmetric system is always better than that of the symmetric system if the performance of the big core in the asymmetric system is the same as that of the core in the symmetric system. Therefore, we consider preserving one core of RISC-V GPGPU based on symmetric architecture and transforming the remaining cores into smaller cores. Taking Inspur's F10A FPGA board as an example, it can accommodate up to 128 basic hardware units. Under the design of the symmetric architecture, there are 8 cores of the same size, each of which occupies 16 basic hardware units. On the premise that the total number of basic hardware units is unchanged, it is modified into an asymmetric architecture with one large core, which occupies 16 basic hardware units, and 112 small cores, each of which occupies 1 basic hardware unit. Under the condition that the proportion of parallel tasks is 0.9, the relative reduction is $\frac{\Delta T}{T_{sym}} = 38.3\%$ for asymmetric multicore architecture and symmetric multicore architecture.

5.6.3.4 Software/Hardware Interface Optimization

In this subsection, we discussed the implementation of the Smart_xPU prototype platform, which serves as our internal design verification and architectural exploration tool. By leveraging innovative compiler technologies and selecting high-level abstraction models, we aimed to reduce design time and improve the quality of computer chips and platforms. Wherever possible, we used reliable open-source toolkits in our prototype design flow.

The Smart_xPU platform is a lightweight, distributed heterogeneous architecture that features an x86 host and multiple xPU devices. It uses an LLVM-layered compilation solution, where LLVM (Low Level Virtual Machine) provides modular and reusable compiler and toolchain technologies for building both compiler front ends and back ends. On the x86 host side, we used the standard GCC compiler, while the xPU hardware was implemented with open-source OpenCL and RISC-V. The RISC-V GNU compiler toolchain was also used for developing xPU device software.

Open Computing Language (OpenCL) was the first open, royalty-free programming standard designed for general-purpose parallel programming across heterogeneous computing systems, including single- and multicore CPUs, GPUs, FPGAs, and digital signal processors (DSPs). It offers a unified programming environment, enabling developers to write efficient, lightweight code for HPC servers, desktops, and handheld devices. Originally proposed by Apple in 2008, OpenCL was later enhanced through collaborations with AMD, IBM, Intel, and Nvidia before being managed by the Khronos Group [37]. Figure 5.17 provides a comparison between the OpenCL programming paradigm and the traditional one.

Several silicon vendors, including ARM, Intel, AMD, IBM, and Nvidia [38], have adopted OpenCL in their designs, either directly or through abstraction models layered on top of OpenCL. Open-source implementations such as Portable Computing Language (PoCL) [39] are also available, as well as OpenCL-based abstraction models like Khronos' SYCL [40] and Intel's DPC++ [41]. Additionally,

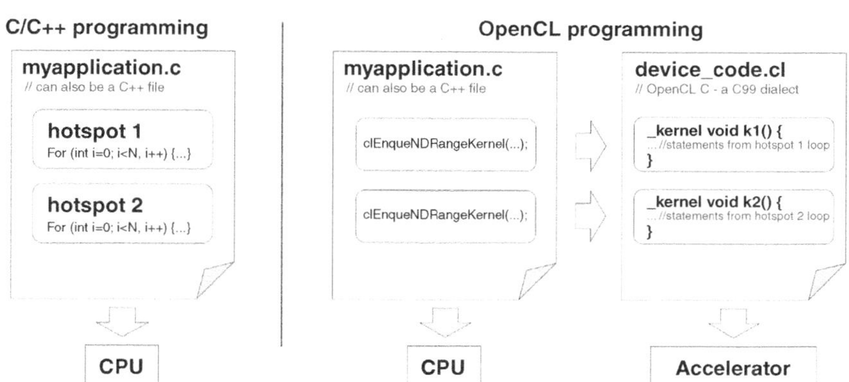

Figure 5.17 Comparison of the OpenCL programming paradigm and the traditional programming paradigm. Source: Adapted from [38, 47].

open-source architectures based on OpenCL include Nyuzi GPGPU [42], NVIDIA [43], Vortex GPGPU [44], OpenVGA [45], and MIAOW GPU [46].

The Smart_xPU architecture differentiates itself from other heterogeneous acceleration platforms through a unique and efficient API for direct communication between the chip kernel engine and the communication requirements initiated by the engine. This programming interface allows for seamless chip-level communication, making Smart_xPU distinct from other platforms. The overall software architecture is illustrated in Figure 5.18.

In the Smart_xPU architecture, the xPU life cycle is closely integrated with the ICMP hardware, the DHCP module, and the DHCP software architecture. Once the xPU-to-xPU over Ethernet-IP-Ethernet channel is established, each xPU core

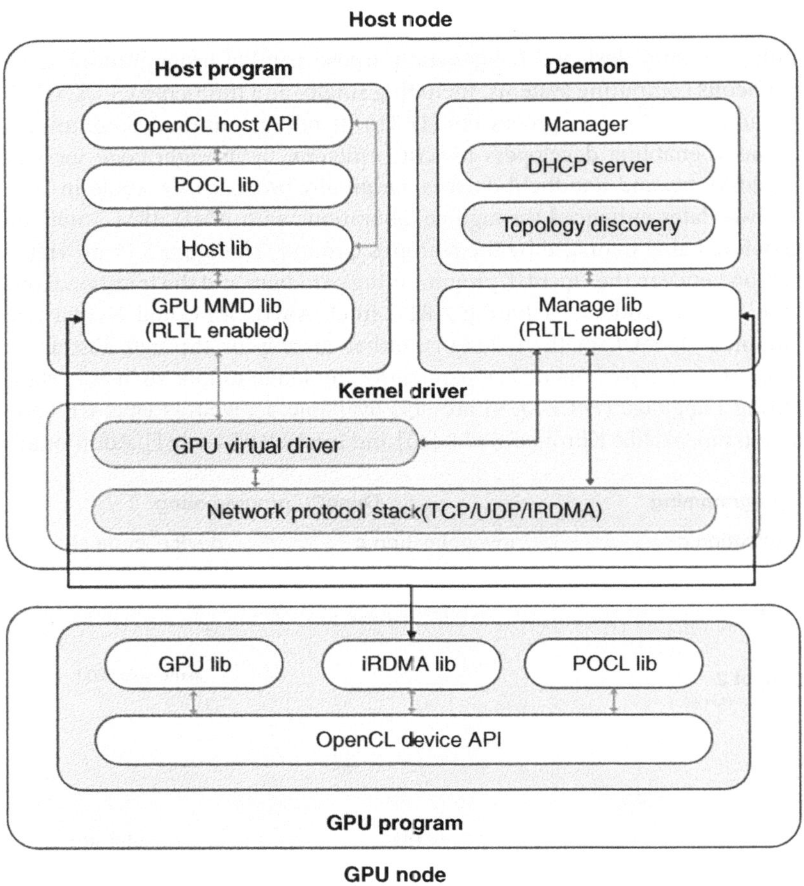

Figure 5.18 The Smart_xPU software architecture overview.

operates autonomously as a peer-to-peer computing and communication node in the network, no longer relying on a host node connected via PCIe.

This Ethernet-based interconnection and communication technology plays a critical role in enabling large-scale, agile deployments. By driving the architecture away from traditional CPU-centric designs and towards an autonomous xPU/CPU dual-engine architecture, the Smart_xPU platform sets the stage for the future of computing architectures.

5.6.3.5 Analyzing the Smart_xPU Architecture

As previously discussed, the "dark silicon" phenomenon is a major challenge facing all high-performance computing (HPC) chips today. Key issues include excessive power and area consumption in processor chips, limitations in silicon VLSI processes, and increasing interdependencies within software stacks, all contributing to various architectural performance bottlenecks. At the 2021 ASPLOS conference, David Patterson and Chris Lattner underscored the urgent need for significant changes in traditional computer architecture and the co-design of hardware and software to effectively address these challenges.

In response, the Smart_xPU architecture offers two crucial contributions for the CNC era:

Verification Platform: The Smart_xPU architecture provides a platform for CNC research, helping to mitigate the effects of "dark silicon" through simultaneous optimization of both software and hardware. By streamlining on-chip and off-chip data transfers, enabling multiple xPU cores to function autonomously, and integrating seamlessly with traditional architecture, the Smart_xPU prototype paves the way for system architects to transition from CPU-centric models to a dual-engine architecture that combines CPUs and autonomous xPUs.

Architectural Exploration Platform: The Smart_xPU architecture also serves as a foundation for the next generation of distributed heterogeneous server pool platforms. Unlike traditional network architectures with fixed network functions, this architecture is built on standard Ethernet and is designed to significantly reduce power consumption and latency, aligning with global "dual carbon" goals for sustainable, green deployment.

The Smart_xPU software and hardware stacks focus on key areas such as energy-efficient data paths, convergence of computing network interfaces, parallel xPU computing, and compatibility with mainstream programming standards.

The Smart_xPU prototype is designed for academic researchers to explore and verify its implementation. An agile development approach is employed, allowing iterative exploration of both hardware and software architectures. The implementation utilizes FPGA devices for flexibility, specifically leveraging the commercial INSPUR F10A FPGA accelerator.

The complete software and hardware stacks are open-source to facilitate easy adoption and verification by researchers across academia, government, and industry. The software programming layer adheres to the OpenCL standard, using the POCL open-source framework, while the accelerator architecture is based on the open-source RISC-V architecture, along with a corresponding development and compilation environment.

To support a "single network card" capable of handling all future configurable network functions, the architecture is designed to be independent of any specific technology process. Protocols such as PCIe over Ethernet, IP, ICMP, and DHCP are redesigned for compatibility with both legacy and future systems.

The Smart_xPU architecture also integrates a convergence engine for computing network interfaces and an API-driven parsing engine for autonomous communication. This ensures autonomous operation of the xPU architecture, reducing reliance on any specific CPU. The xPU over Ethernet implementation is carefully partitioned between CPU and FPGA, enabling a parallel, pipelined design on both sides.

Figure 5.19 illustrates the Smart_xPU architecture, which unifies these user requirements and technical analyses. To maintain backward compatibility, the Smart_xPU communicates with traditional host nodes through NICs. The entire distributed platform is implemented on the Wave F10A FPGA accelerator, with the software programming layer built on the OpenCL standard and the POCL framework. The xPU cores are implemented on FPGA using an open-source RISC-V design.

The software stack is split between the x86 host and the xPU, utilizing an LLVM-layered compilation approach. The x86 host employs GCC to compile host programs, while the RISC-V side uses the RISC-V GNU Compiler Toolchain for device programs. To optimize resource use, the xPU architecture employs a large core for sequential tasks and smaller cores for parallel tasks, enhancing both FPGA utilization and overall system performance. Additionally, hardware blocks like the Ethernet MAC layer and RDMA engines are implemented in FPGA hardware.

5.6.4 Summary of the Smart_xPU Architecture

Section 5.6 introduces an emerging distributed parallel architecture called Smart_xPU, along with a lightweight, cost-effective prototype platform for verification and architectural exploration in the context of the CNC era. From an architectural standpoint, Smart_xPU aims to leverage three types of parallelism: instruction-level parallelism (ILP), data-level parallelism (DLP), and thread-level parallelism (TLP). Currently, the existing Smart_xPU implementation is not fully optimized, leading to over-configurations and under-utilization of resources.

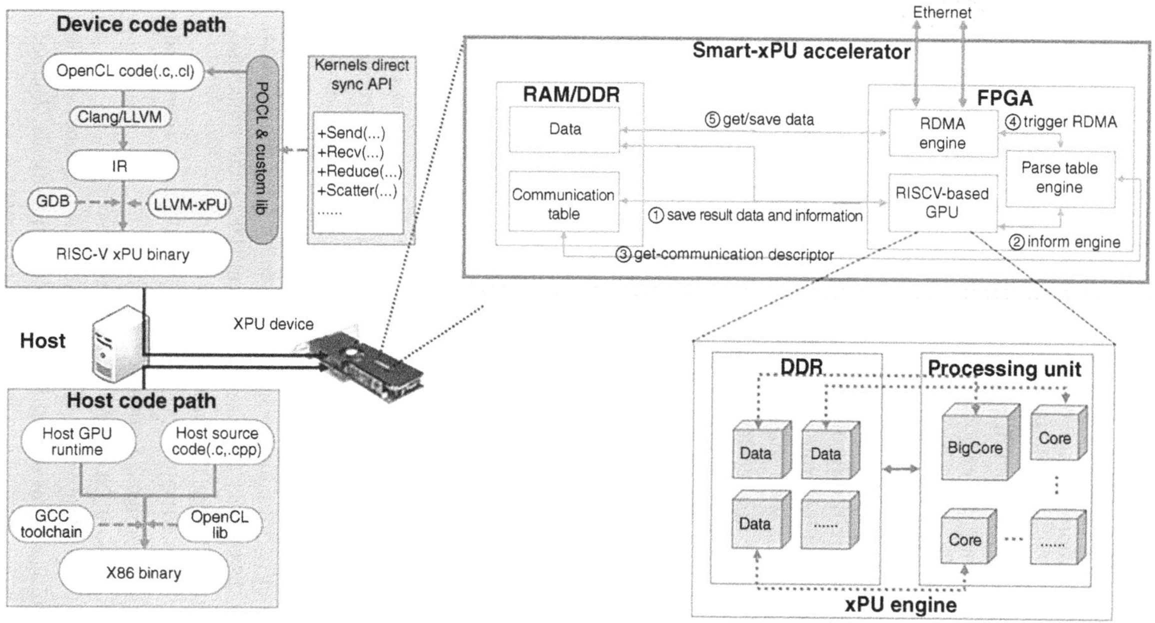

Figure 5.19 The Smart_xPU prototype architecture.

Future improvements will focus on optimizing the chip architecture, compilers, runtime libraries, kernel drivers, and establishing interoperability standards, opening up new avenues for research.

The traditional boundaries of computing architecture, particularly around scale-up expansion, have become increasingly blurred. Memory, I/O expansion, and high-speed interconnects are no longer constrained by conventional resource access limitations, making it challenging to differentiate between von Neumann, Harvard, and non-von Neumann parallel architectures. However, the concept of distributed acceleration server pools has gained broad recognition in the industry. As a result, the Smart_xPU software and hardware stack presents an opportunity for academia, industry, and the research community to address many of the pain points in CNC systems.

In summary, the architectural design of Smart_xPU represents a distributed computing engine that integrates computation and networking. It aligns user and technical requirements with the convergence of computing networks, offering a valuable reference for further design and implementation. The next chapter will delve into an end-to-end 5G network example to explore the network functions (NFs) and their interfaces, as NFs are typically implemented as software running on servers.

References

1 International Telecommunication Union (ITU) Home Page. ITU. 2024. https:/itu.int/ (accessed on 1 September 2024).

2 IMT Vision – Framework and overall objectives of the future development of IMT for 2020 and beyond. ITU September 2015. https://www.itu.int/dms_pubrec/itu-r/rec/m/R-REC-M.2083-0-201509-I!!PDF-E.pdf (accessed on 1 September 2024).

3 Introducing 3GPP. The Third Generation Partnership Project (3GPP). https://3gpp.org/about-us/introducing-3gpp (accessed on 1 September 2024).

4 3GPP (2022). TS38.101-1 NR User Equipment (UE) radio transmission and reception, Part 1: Range 1 Standalone.

5 3GPP (2022). TS38.101-2 NR User Equipment (UE) radio transmission and reception, Part 2: Range 2 Standalone.

6 3GPP (2017). TS38.401 NG-RAN, Architecture description.

7 Choudhary, U. and Janyani, V. (2021). Orthogonal frequency division multiplexing for mitigation of optical fiber channel impairments. *Journal of Optics* 50: 186–200. https://doi.org/10.1007/s12596-021-00697-9.

8 3GPP (2022). TS38.300 NR, NR and NG-RAN Overall Description, Stage 2.

9 3GPP (2017). TS23.501 System Architecture for the 5G System, Stage 2.

10 "Computing-network convergence empowers the digital economy, unblocking the 'information arteries'." China News Service (CNS), 5 November 2022. https://baijiahao.baidu.com/s?id=1748662470365660521&wfr=spider&for=pc (accessed on 1 September 2024).

11 "Achieving China's Dual Carbon Goals." Shanghai Observer, 5 June 2022. https://export.shobserver.com/baijiahao/html/494900.html (accessed on 1 September 2024).

12 Zhao, Y., Lu, J., and Chen, X. (2020). A dynamically reconfigurable accelerator design using a sparse-Winograd decomposition algorithm for CNNs. *Computers, Materials and Continua* 66 (1): 517–535.

13 Ghaffari, A. and Savaria, Y. (2020). CNN2Gate: toward designing a general framework for implementation of convolutional neural networks on FPGA. Published in arXiv.org on 6 April 2020. Corpus ID: 215548724.

14 Véstias, M.P., Rui, P.D., Sousa, J. et al. (2020). A fast and scalable architecture to run convolutional neural networks in low density FPGAs. *Microprocessors and Microsystems* 77: 103136.

15 Fowers, J., Ovtcharov, K., Papamichael, M.K. et al. (2019). Inside project brainwave's cloud-scale, real-time AI processor. *IEEE Micro* 39 (3): 20–28.

16 Chung, E., Fowers, J., Ovtcharov, K. et al. (2018). Serving DNNs in real time at datacenter scale with project brainwave. *IEEE Micro* 38 (2): 8–20.

17 Shainer, G., Ayoub, A., Lui, P. et al. (2011). The development of Mellanox/NVIDIA GPUDirect over InfiniBand – a new model for GPU to GPU communications. *Computer Science-Research and Development* 26: 267–273.

18 Ammendola, R., Bernaschi, M., Biagioni, A. et al. (2013). GPU peer-to-peer techniques applied to a cluster interconnect. *2013 IEEE International Symposium on Parallel & Distributed Processing, Workshops and PhD Forum*, pp. 806–815. 20–24 May 2013, Boston MA USA. IEEE.

19 Agostini, E., Rossetti, D., and Potluri, S. (2017). Offloading communication control logic in GPU accelerated applications. *2017 17th IEEE/ACM International Symposium on Cluster, Cloud and Grid Computing (CCGRID)*, pp. 248–257. 14–17 May 2017. Madrid Spain, IEEE.

20 Agostini, E., Rossetti, D., and Potluri, S. (2018). GPUDirect Async: exploring GPU synchronous communication techniques for InfiniBand clusters. *Journal of Parallel and Distributed Computing* 114: 28–45.

21 Oden, L. and Fröning, H. (2013). GGAS: global GPU address spaces for efficient communication in heterogeneous clusters. *2013 IEEE International Conference on Cluster Computing (CLUSTER)*, pp. 1–8. 23–27 September 2013, Indianapolis Indiana USA. IEEE.

22 Daoud, F., Watad, A., and Silberstein, M. (2016). GPUrdma: GPU-side library for high performance networking from GPU kernels. *Proceedings of the 6th*

International Workshop on Runtime and Operating Systems for Supercomputers, pp. 1–8. 1 June 2016, Kyoto Japan. ISBN: 978-1-4503-4387-9.

23 Silberstein, M., Kim, S., Huh, S. et al. (2016). GPUnet: networking abstractions for GPU programs. *ACM Transactions on Computer Systems (TOCS)* 34 (3): 1–31.

24 Weerasinghe, J., Abel, F., Hagleitner, C. et al. (2015). Enabling FPGAs in hyperscale data centers. *2015 IEEE 12th Intl Conf on Ubiquitous Intelligence and Computing and 2015 IEEE 12th Intl Conf on Autonomic and Trusted Computing and 2015 IEEE 15th Intl Conf on Scalable Computing and Communications and Its Associated Workshops (UIC-ATC-ScalCom)*, pp. 1078–1086. 10–14 August 2015, Beijing China. IEEE.

25 Wang, Y., Kan, H., Su, D. et al. (2020). Energy efficient computing offloading mechanism based on FPGA cluster for edge cloud. *EITCE 2020: 2020 4th International Conference on Electronic Information Technology and Computer Engineering*. 6–8 November 2020, Xiamen China, pp. 1120-1125.

26 Li, R., Dongdong, S., and Kan, H. (2021). A fast, scalable, and energy-efficient edge acceleration architecture based on FPGA cluster. *17th ACM International Conference on Emerging Networking EXperiments and Technologies, CoNEXT 2021*. 7–10 December 2021, pp. 479–80. Virtual Conference Germany. Association for Computing Machinery, Inc.

27 Balle, S.M., Tetreault, M., and Dicecco, R. (2020). Inter-kernel links for direct inter-FPGA communication. *Intel White Paper (WP-01305–1.0)*.

28 Kan, H., Li, R., Su, D. et al. (2020). Trusted edge cloud computing mechanism based on FPGA cluster. *2020 IEEE 8th International Conference on Computer Science and Network Technology (ICCSNT)*, pp. 146–149. 20–22 November 2020, Dalian China. https://doi.org/10.1109/ICCSNT50940.2020.9304996.

29 Duan, X., Yao, H., Yuexia, F. et al. (2021). Computing force network technologies for computing and network integration evolution. *Telecommunications Science* 37 (10): 76–85.

30 Shumarayev, S. (2017). Heterogeneous Modular Platform. Hot Chips: A Symposium on High Performance Chips, HC29, Cupertino, CA, USA, 20–22 August 2017. https://old.hotchips.org/wp-content/uploads/hc_archives/hc29/HC29.22-Tuesday-Pub/HC29.22.50-FPGA-Pub/HC29.22.523-Hetro-Mod-Platform-Shumanrayev-Intel-Final.pdf (accessed on 17 October 2024).

31 Huang, X., Guo, Z., Song, M. et al. (2021). AccelSDP: a reconfigurable accelerator for software data plane based on FPGA SmartNIC. *Electronics* 10 (16): 1927.

32 Yan, G.H. et al. (2021). Technical White Paper of Data Processing Unit, Revision 1.0, October 2021. Chinese Academy of Sciences, Institute of Computing Technology.

33 Cheng, H., Wang, Y., Fan, J. et al. (2021). Research on DC network transmission handover technology based on user mode sharing. *Proceedings of the 5th International Conference on Computer Science and Application Engineering (CSAE '21)*. Association for Computing Machinery, Article 79, pp. 1–5, 19–21 October 2021, Sanya China.

34 Wang, Y., Cheng, H., Fan, J. et al. (2021). Research on high performance transmission technology of DC based on network awareness. *Proceedings of the 5th International Conference on Computer Science and Application Engineering (CSAE '21)*. Association for Computing Machinery, Article 78, pp. 1–6, 19–21 October 2021, Sanya China.

35 Feng, Y. and Deng, Q. (2011). A study of Amdahl's law performance model on asymmetric multicore system. *Microelectronics & Computer* 28 (8): 32–34.

36 Borkar, S. (2007). Thousand core chips: a technology perspective. *Proceedings of the 44th Annual Design Automation Conference*, pp. 746–749, 4–8 June 2007, San Diego CA USA.

37 Stone, J.E., Gohara, D., and Shi, G. (2010). OpenCL: a parallel programming standard for heterogeneous computing systems. *Computing in Science & Engineering* 12 (3): 66–72.

38 NVIDIA Developer Program. NVIDIA Open Computing Language OpenCL Home Page. https://developer.nvidia.com/opencl (accessed on 17 October 2024).

39 Jääskeläinen, P., de La Lama, C.S., Schnetter, E. et al. (2015). pocl: a performance-portable OpenCL implementation. *International Journal of Parallel Programming* 43: 752–785.

40 The Khronos® SYCL Working Group. The SYCL Home Page. https://www.khronos.org/sycl (accessed on 17 October 2024).

41 Reinders, J., Ashbaugh, B., Brodman, J. et al. (2021). *Data Parallel C++: Mastering DPC++ for Programming of Heterogeneous Systems Using C++ and SYCL*. Springer Nature.

42 Bush, J., Taherinejad, N., Willegger E et al. (2021). Nyuzi: an open source GPGPU for graphics, enhanced with OpenCL compiler for calculations. *IEEE Design, Automation & Test in Europe*, 1. IEEE.

43 Zhou, G., Zhou, J., and Lin, H. (2018). Research on NVIDIA deep learning accelerator. *2018 12th IEEE International Conference on Anti-counterfeiting, Security, and Identification (ASID)*, pp. 192–195. 9–11 November 2018, Xiamen China. IEEE.

44 Tine, B., Elsabbagh, F., Yalamarthy, K. et al. (2021). Vortex: extending the RISC-V ISA for GPGPU and 3D-GraphicsResearch. *arXiv preprint*. arXiv:2110.10857.

45 Suggate, P. (2009). OpenVGA: an open-source PCI graphics adapter. Master's thesis, University of Otago.

46 Balasubramanian, R., Gangadhar, V., Guo, Z. et al. (2015). MIAOW – An open source RTL implementation of a GPGPU. *2015 IEEE Symposium in Low-Power and High-Speed Chips (COOL CHIPS XVIII)*, 19–21 April 2023, Yokohama Japan, pp. 1–3. https://doi.org/10.1109/CoolChips.2015.7158663.

47 The Khronos® OpenCL Working Group. The OpenCL™ C Specification, Version V3.0.16, 4 April 2024. https://registry.khronos.org/OpenCL/specs/3.0-unified/pdf/OpenCL_C.pdf (accessed on 17 October 2024).

6

Software Architecture

Having introduced the hardware architecture in the last chapter, we will make use of an end-to-end (E2E) 5G system example [1] in this chapter to illustrate the functionality of the network functions (NFs) and their interfaces. We will also introduce some important concepts behind the 5G software.

6.1 End-to-End Example of 5G System

Here, we introduce an example of an E2E 5G system showing how UEs are connected to data network (DN). See the Figure 6.1.

6.1.1 High-Level Description

The E2E 5G system shown above comprises five key parts. From the end users' side, the wireless connection can be made from mobile user equipment (UE) such as cell phones, tablets, and Internet of Things (IoT) devices to the radio access network (RAN). The RAN communicates with UEs wirelessly via a radio remote head (RRH). The RAN further comprises a distributed unit (DU) and a central unit (CU). Signals received at the RAN can be processed by an Edge Cloud or be passed on to a 5G Core. The 5G Core consists of two main categories, namely, the control plane (CP) and user or data plane (DP), which are responsible for the control and management functions and for transferring the control signals, user information, and data across the 5G system. The 5G Core is connected to applications and services in a DN which can be the Internet, IP multimedia subsystem (IMS), or roaming service providers. We will look at the key parts of the 5G network, from RAN to applications and services in the following subsections.

Edge Computing Acceleration: From 5G to 6G and Beyond, First Edition.
Patrick Hung, Hongwei Kan, and Greg Knopf.

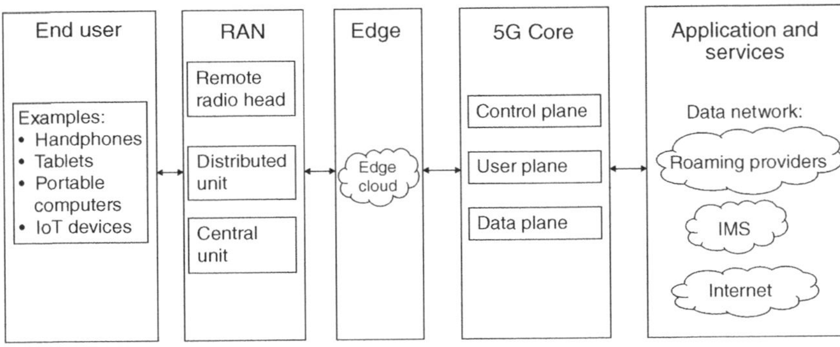

Figure 6.1 Example of an end-to-end 5G system. Source: Adapted from [1].

6.1.1.1 5G Radio Access Network

The key function of a RAN is to connect UEs wirelessly to the 5G Core network. To achieve this goal, an RAN has several characteristics, such as the ability to handle high-frequency bands in 5G. As high-frequency signals attenuate over long distances as they propagate through the air, more cells and antenna techniques are needed in deploying a 5G network. Other 5G RAN characteristics include the capability to support a large number of devices in mMTC (massive machine type communications) scenarios and real-time applications, which requires URLLC, which further calls for the ability to support network slicing and mobile edge computing. In this section, we will briefly cover the three components of a 5G RAN.

Remote Radio Head (RRH) The RRH is responsible for handling radio communications between UEs and the RAN. It processes radio signals before transmission to UEs and receives radio signals from UEs. Equipped with antennas, the RRH uses MIMO technology as discussed in the previous chapter, to target a specific UE while minimizing interferences to other UEs.

Distributed Unit (DU) Deployed between the RRH and the Central Unit (CU), the DU's key responsibility is performing signal processing in the RAN for both radio signal transmission and recovery of baseband signal. The DU achieves both tasks through digital signal processing techniques such as modulation or demodulation, channel estimation, error correction, and beamforming.

Central Unit (CU) The CU controls multiple DUs in the RAN through scheduling, radio resource management, and network management. The key goal is to ensure E2E connectivity and good service quality delivered to users by optimizing radio resources in the RAN. It manages radio resources through spectrum allocation, power control, and interference management. It distributes the 5G Core data to specific DUs and conducts handover between DUs when a UE moves from one DU area to another. It monitors the network conditions and conducts the traffic routing.

6.1.1.2 Edge

Deployed between the RAN and the 5G Core lies a 5G Edge, which performs processing and storing of data at the network edge. Since the 5G Edge is close to end users, data does not need to travel through the DN. Processing of data at the edge enables short response time and low latency, which are critical for real-time applications such as video streaming, virtual reality, and augmented reality.

The 5G Edge comprises computing resources capable of handling a large amount of data from many IoT devices. This arrangement facilitates smart factory or autonomous vehicle applications in which a high volume of data is generated and can be processed and analyzed quickly so that appropriate action can be triggered (e.g., alert or control) in a real-time manner. By processing IoT data at the edge, heavy traffic in the 5G network is avoided.

One important component in the 5G Edge is the hypervisor, also known as a virtual machine (VM) monitor. It is a software layer that enables the creation and management of VMs on a physical server. It allows multiple VMs to run on the same physical machine by dividing the available hardware resources, such as CPU, memory, storage, and network bandwidth, among the VMs.

In the 5G Edge, the hypervisor creates and manages the virtualized infrastructure that supports the deployment of edge services and applications. It provides a layer of abstraction between the physical infrastructure and the VMs, enabling the efficient utilization of resources, supporting dynamic allocation, and scaling of resources based on the needs of the edge applications.

The hypervisor also provides security and isolation for the VMs running on the edge network. Each VM runs in its own isolated environment, which prevents one VM from accessing or interfering with the resources or data of another VM. This is important in 5G Edge, where multiple applications and users may share the same physical infrastructure.

Moreover, the hypervisor supports the migration of VMs between physical servers, which enables workload balancing and resiliency for the edge network. If a physical server fails, the VMs running on that server can be quickly migrated to another server without disrupting the running applications.

6.1.1.3 5G Core

Two key areas exist in 5G: the CP and the DP. We will discuss each of these areas, starting with the CP. Note that there are other functions that the following network elements perform. Details of these other functions are available in [2].

Control Plane CP is responsible for controlling the 5G system. It has several different functions. A high-level description of the key functions is as follows:

Access and Mobility Management Function (AMF) is mainly responsible for network signaling and security in access control and mobility management. AMF allows authorized UEs to access the 5G network and keeps track of the base stations serving them. It also ensures that connections are maintained when a UE moves from one base station to another and handles idle mobility.

Session Management Function (SMF) is responsible for establishing, modifying, and releasing sessions. It starts by allocating an IP address to an authorized UE (by AMF). It directly communicates with a UPF to maintain a particular session between the UE and the DN. It also selects and controls user plane (UP) functions.

Policy Control Function (PCF) governs network behavior by providing and executing rules to the CP functions, such as rules for QoS control and charging.

Network Exposure Function (NEF) exposes capabilities and events of NFs to other network elements, such as third party, application functions (AFs), and edge computing.

NF Repository Function (NRF) discovers services and transfers information of the discovered services to NF instances.

Unified Data Management (UDM) manages user identification (e.g., generation of user authentication credentials) and access authorization based on subscription data.

Authentication Server Function (AUSF) authenticates UE.

Network Slice Selection Function (NSSF) manages the selection of appropriate network slices to serve UEs.

Short Message Service Function (SMSF) supports the SMS transfer over NAS.

Location Management Function (LMF) tracks location information for a UE.

User or Data Plane (DP) The key component in the DP is the UP function (UPF), which realizes the concept of CP and UP separation (CUPS). It is mainly used to transfer user data between the RAN and DN elements such as the Internet. The UPF comprises the following key features:

(a) Handling of user data packet (PDU) for functions such as packet routing and QoS for the UP.
(b) **Mobility Anchoring:** communicating with the SMF of the CP.

Another key function of the DP is managing the data transfer between the 5G Core and the other network components. There are two components here. First, a unified data repository (UDR) supports the storage and recovery of data such as identity, subscription details, device information, and usage data in UDM and PCF, which facilitates services, network resources management, and securing services. Second, unstructured data storage network function (UDSF) provides an optional function for the storage and recovery of unstructured data in any NFs.

6.1.1.4 Application and Services

Applications and services in a DN comprise the Internet, IP multimedia subsystem (IMS), and roaming providers. In this subsection, we focus on the IMS.

IMS is a standardized architectural framework that enables various multimedia services to be delivered in IP networks. The framework integrates multimedia services such as voice calls, video calls, conferencing, and text messaging into one platform that users using different devices, applications, and networks can use. The IMS standard enables interoperability for various networks.

The key elements of an IMS include the call session control function (CSCF), home subscriber server (HSS), application server (AS), and media resource function (MRF). These elements provide functions such as call session management between terminals, call routing within a network, user data storage, such as user profile, authentication information and service subscription information, provision of various multimedia services, and media processing services.

IMS provides flexibility not only for network operators but also for users due to the various multimedia services offered. It is also scalable, as it can handle several devices and applications across many networks.

6.1.2 Interfaces

The interfaces between different components are summarized as follows.

6.1.2.1 N1: Between 5G Core and User Equipment

This is a logical interface between the UE and the control plan communication with the AMF on the control plan. It is responsible for establishing, maintaining, and releasing connections between the UE and the network.

Key functions of the N1 interface include:

- Authentication and authorization of the UE.
- Mobility management, including tracking and paging of the UE.
- Quality of Service (QoS) management.
- Session management and control of data transfer.
- Support for network slicing.

6.1.2.2 N2: Between 5G Core and Base Station

It provides a connection between the RAN and AMF on the CP. A key function of the N2 interface is to support mobility management, allowing the 5G network to manage user movements between different base stations. It also facilitates session management, allowing the 5G network to establish, modify, and release data sessions between the UE and the 5G Core network.

The N2 interface also offers flexibility and scalability, allowing an RAN to work with various network needs and requirements. It can support different transport protocols, including Ethernet, IP, and MPLS.

6.1.2.3 N3: Between RAN and User Plane Function

It provides a link transferring user data from a RAN to the UPF.

6.1.2.4 Other Interfaces Include the Following

- **N4:** between the SMF (Session Management Function) and the UPF (User Plane Function).
- **N5:** between the PCF (Policy Control Function) and AF (Application Function).
- **N6:** between the UPF (User Plane Function) and a DN (Data Network).
- **N7:** between the SMF (Session Management Function) and the PCF (Policy Control Function).
- **N7r:** between V-PCF (Visited Policy Control Function) and H-PCF (Home Policy Control Function).
- **N8:** between UDM (Unified Data Management) and AMF (Access and Mobility Management Function).
- **N9:** between two Core UPFs (User Plane Functions).
- **N10:** between UDM (Unified Data Management) and SMF (Session Management Function).
- **N11:** between AMF (Access and Mobility Management Function) and SMF (Session Management Function).
- **N12:** between AMF (Access and Mobility Management Function) and AUSF (Authentication Server Function).
- **N13:** between UDM (Unified Data Management) and AUSF (Authentication Server Function).
- **N14:** between two AMFs (Access and Mobility Management Functions).
- **N15:** between the PCF (Policy Control Function) and the AMF (Access and Mobility Management Function) in case of non-roaming scenario and V-PCF (Visited Policy Control Function) and AMF (Access and Mobility Management Function) in case of roaming scenario
- **N16:** between two SMFs (Session Management Functions) in roaming case between V-SMF (Visited Session Management Function) and the H-SMF (Home Session Management Function).
- **N22:** between AMF (Access and Mobility Management Function) and NSSF (Network Slice Selection Function).

6.2 Network Slicing Architecture, Software-Defined Network, and Network Function Virtualization

Traditionally, NFs are implemented as dedicated hardware, which imposed limitations on network expansion or connectivity with other network elements. Network slicing architecture, software-defined networking (SDN), and network function virtualization (NFV) are all related concepts that together enable the creation of agile, flexible, and programmable modern networks capable of supporting diverse use cases and applications. In this section, we will cover these concepts.

6.2.1 Network Slicing Architecture

A 5G network is an integrated computing and communications network. Edge computing defines the 5G computing architecture, and network slicing defines the 5G communications network.

Network slicing – also known as enhanced virtual private network – is an emerging 5G-network paradigm leveraging network softwarization and network virtualization to enable the instantiation of multiple virtual networks (also known as network slices) over the same physical network. These mutually isolated E2E network slices encompass compute, storage, and networking functions and can be created on demand.

Unlike traditional VPN technology, network operators can allocate dedicated compute storage resources and customized functions to each network slice. This allows each slice to always meet its specific cost, latency, bandwidth, and security levels at all times. The ultimate objective is to improve system scalability and flexibility while minimizing capital and operating expenses (OPEX) (Figure 6.2).

Network slicing is based on two underlying technologies:

- **Software-Defined Network (SDN)** reduces the difficulties to maintain complicated 5G network. It also helps improve system scalability, flexibility, and reliability.
- **Network Function Virtualization (NFV)** helps reduce capital expenses (CAPEX) and OPEX. It also allows users to easily create, delete, or reconfigure remote network functions.

Open network automation platform (ONAP; www.onap.org), which was established in February 2017, has been developing an SDN/NFV-based network platform, laying the groundwork for the implementation of 5G network slicing.

The ONAP project addresses the rising need for a common automation platform for telecommunication, cable, and cloud service providers – and their solution

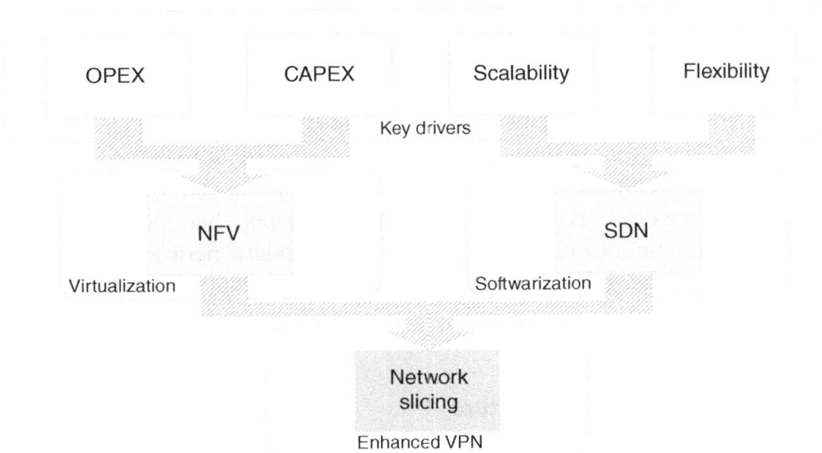

Figure 6.2 Network slicing architecture.

providers – to deliver differentiated network services on demand, profitably, and competitively, while leveraging existing investments.

6.2.1.1 Software-Defined Network (SDN)

Historically, the control and data elements in networking equipment were tightly coupled, providing many proprietary solutions. Unfortunately, as network functions get more complicated, this tightly coupled architecture has become very inflexible and expensive to maintain.

SDN has emerged as a new networking paradigm, which allows 5G network providers to easily control a wide spectrum of heterogeneous network traffic and to manage a wide range of networking equipment from multiple vendors.

In 2004 the Internet Engineering Task Force (IETF) began considering various ways to decouple the control functions and data switches in networking. The computer networking research group at Stanford University was instrumental in developing the OpenFlow protocol, a key SDN building block. The Open Network Foundation (ONF) was founded in 2011 to promote SDN and manage the OpenFlow standard.

The SDN architecture contains three different layers, known as planes (Figure 6.3):

- **Data plane** contains the configurable network devices in a domain.
- **Control plane** is a computing layer – typically using commercial off-the-shelf (COTS) products – tasked to control and configure the DP.
- **Management plane** contains several network management applications that use network protocols (e.g., HTTP) to manage the CP.

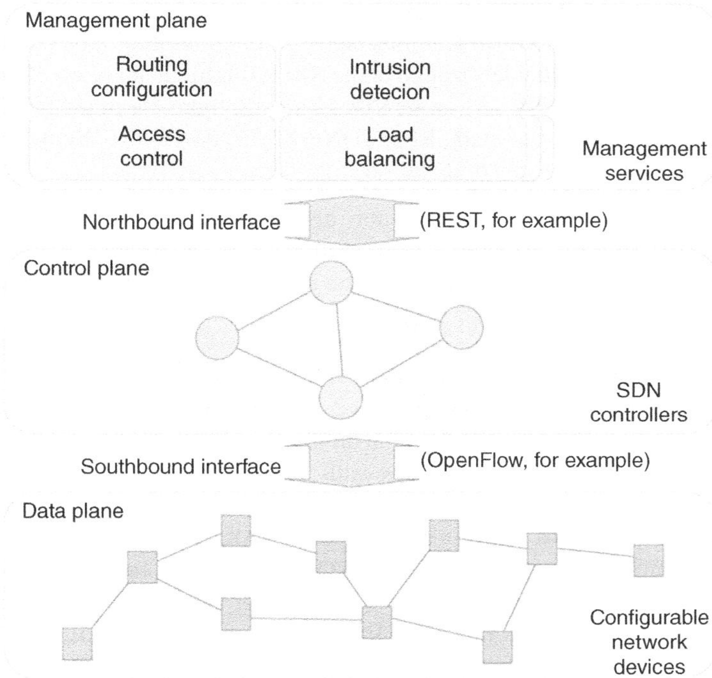

Figure 6.3 SDN architecture.

SDN has received huge commercial successes in the past decade. According to the market research firm MarketsandMarkets, the global SDN market has grown from USD 0.5 billion in 2015 to USD 13.7 billion in 2020 and will reach USD 32.7 billion by 2025 at a Compound Annual Growth Rate (CAGR) of 19.0%.

The leading SDN vendors include Cisco, Dell EMC, HPE, Vmware, Huawei, Juniper Networks, Nokia, Oracle, Citrix, Extreme Networks, Infovista, NEC, Pluribus Networks, Arista Networks, CloudGenix, Cumulus Networks, DataCore Software, Fortinet, Fujitsu, HiveIO, Lenovo, NetApp, Pica8, Pivot3, and Scale Computing.

The SDN architecture contains two main interfaces:

- **Southbound Interface (SBI):** The CP and DP are connected by an SBI, which provides an abstraction of network device hardware functionality. The SBI allows the CP to push configuration information and install flow entries in the DP. OpenFlow is the de facto SBI; however, there are also some other alternatives, including ForCES, OpFlex, and NetConf.

- **Northbound Interface (NBI):** The CP and management plane are connected by a NBI, which provides an abstraction for network programming. Currently, unlike the SBI, there is no standardization of the NBI at this moment. Many SDN controllers can support REST API as an NBI. However, there are many other NBIs, including Java RPC, Quantem, RESTCONF, XMPP, NETCONF, Neutron, Onix API, PANE API, and some other ad-hoc interfaces.

In addition to these NBI and SBI, SDN may also contain **eastbound interface** (EBI) and **westbound interface** (WBI). EBI refers to the interconnection between two SDN domains and WBI refers to the interconnection between SDN and traditional network domains.

6.2.1.2 Network Function Virtualization (NFV)

NFV – used in conjunction with SDN. It allows users to create virtual network functions (VNFs) at a remote network location on-demand. There are many kinds of standard or custom VNFs, including Firewalls, Routers, Compression, Load Balancers, Overlay Tunnels, Deep Packet Inspection (DPI), Intrusion Detection Systems (IDS), and Wide Area Application Services (WAAS).

In 2012, seven mobile network operators (MNOs) at European Telecommunications Standards Institute (ETSI) published a white paper based on SDN, which led to the creation of the Network Functions Virtualization Industry Specification Group (ISG NFV). ISG NFV was founded to drive and promote NFV specifications and technologies (Table 6.1).

Network slicing provides a network-as-a-service (NaaS) model, flexibly allocating and reallocating resources according to dynamic demands, such that it can customize network slices for diverse and complex 5G communication scenarios. Network slicing can provide logical networks with better performance than one-size-fits-all networks.

- A network slice can scale up or down as service requirements as the number of users changes.
- Network slices can isolate the network resources of one service from the others; the configurations among various slices do not affect one another. As such, the reliability and security of each slice can be enhanced. A network slice is customized according to service requirements, which can optimize the allocation and use of physical network resources.

According to MarketsandMarkets, the NFV market size is projected to grow from USD 12.9 billion in 2019 to USD 36.3 billion by 2024 at a Compound Annual Growth Rate (CAGR) of 22.9%. The major factors driving the growth of the NFV market include the demand for network virtualization and automation, the

Table 6.1 Timeline of NFV release.

Date	NFV release	Summary
2013–2014	NFV release 1	The key objectives of NFV Release 1 were to (i) drive convergence on network operator requirements, (ii) include existing standards wherever applicable, and (iii) develop new technical requirements.
2015–2016	NFV release 2	NFV Release 2 covered the common specification stages of requirements, architecture, interfaces, and information models and protocols all the way through to the specification of test cases and suite.
2017–2018	NFV release 3	NFV Release 3 has focused on enriching the NFV Architectural Framework to make NFV ready for global deployment and operations. The set of new features includes (i) support for edge computing and network slicing, (ii) new operational aspects, such as multiple administrative domains and software updates, and (iii) advances in virtualization, such as cloud native VNFs and acceleration technologies.
2019–2020	NFV release 4	NFV Release 4 specification work has started in summer 2019, focusing on (i) NFVI evolution, (ii) enhancing NFV automation and capabilities, (iii) evolving the NFV_MANO (Management and Orchestration) framework, (iv) enhanced reliability and availability, (v) accompanying operationalization, and (vi) security hardening.

increased emergence of cloud services and server virtualization, and the need for advanced network management systems.

Major NFV vendors include Cisco, Ericsson, Huawei, VMware, Nokia, HPE, Dell EMC, Juniper Networks, Affirmed Networks, NETSCOUT, NEC, Ribbon Communications, ZTE Corporation, Fujitsu, Ciena, ECI Telecom, Metaswitch, Mavenir, Radisys, and Wind River.

In principle, there are two main virtualization technologies that could be used to attain device virtualization:

- **Virtual machine** (VM)-based virtualization and
- **Container** (CNT)-based virtualization.

Containers are very lightweight and can be deployed significantly quicker than VMs. As a result, the physical resources can easily be modified in real time. On the other hand, container-based virtualization does not provide very strong isolation between applications compared to VM-based virtualization. We will look at different acceleration methods in the next section.

6.3 Software Acceleration

In addition to hardware acceleration methods discussed in the last chapter, network performance can also be improved by software methods. In this section, we will discuss the key methods available. Although these methods were not specified for 5G, they have been subsequently applied in 5G.

6.3.1 User Space Approach

6.3.1.1 Data Plane Development Kit (DPDK)

DPDK is an open-source project [3] that provides a set of libraries and drivers for accelerating data packet processing for various networking applications (i.e., not specifically for 5G). It was originally developed by Intel but is now managed by the Linux Foundation [4] and supported by a large community of contributors.

DPDK uses a user-space driver model via polling mode drivers (PMDs), which allows applications to bypass the kernel and directly access the NIC hardware. This provides a more efficient and scalable approach to packet processing compared to traditional kernel-based networking stacks. PMDs provide a standardized interface that enables applications to interact with various types of NICs and hardware accelerators. PMDs abstract the underlying hardware details and provide a unified API that applications can use to send and receive network packets. Using DPDK, applications can achieve higher packet processing rates and lower latency, which are important for 5G applications like virtualized network slicing, edge computing, and ultra-reliable low-latency communications (URLLC) [5, 6].

DPDK includes several libraries and tools that can be used to build custom DP applications. For example, the DPDK packet framework (DPDK-PF) provides a set of APIs for packet processing, while the DPDK event framework (DPDK-EF) provides a scalable event-driven architecture for managing packet flows.

DPDK applies a "Zero copy" approach in which no data is copied between kernel space and user space. This approach improves the data transfer efficiency between the network interface card (NIC) and the application running on the server. Typically, when a packet arrives at the NIC, it is copied into a buffer in the kernel space and then again copied into a buffer in the user space before being processed by the application. This double-copy operation can be time consuming and reduces the system's overall throughput.

DPDK provides a mechanism for the application to directly access the buffers in the kernel space without the need for a copy operation. This is achieved by mapping the physical memory of the NIC's buffer into the virtual address space of the application using memory mapping techniques. The application can then read or write data directly to the buffer without any copy operation. To handle large data storage, DPDK uses huge pages, which are big blocks of physical and

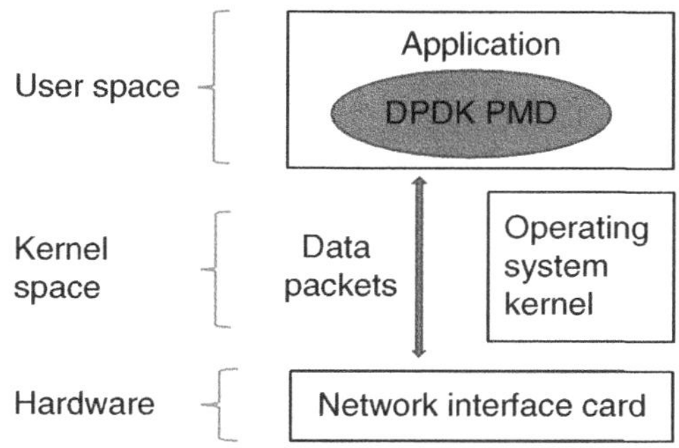

Figure 6.4 DPDK-assisted user space approach.

virtual memory to improve processing performance. The DPDK arrangement is shown in Figure 6.4.

To enable the use of DPDK, hardware support through an NIC is required. Many vendors other than Intel offer NIC support such as AMD, CISCO, Mellanox Technologies, and NXP. Supported NICs refer to hardware for which DPDK provides drivers. Refer to [3] for a list of supported hardware. Recent research has found that it is possible to use NICs that are not originally supported by DPDK libraries to achieve network acceleration [7].

Use of DPDK in Practice In practice, DPDK can be used with other network components (both hardware and software). For instance, Open vSwitch (OVS) [8] is commonly used in a virtualized network function environment. Open vSwitch (OVS) is a software-based virtual switch that allows network administrators to manage network traffic between VMs on a server. Originally not intended for 5G applications, in which a large amount of data is required, OVS uses the kernel space approach for handling data packets. In the 5G networks, OVS can be used to manage network traffic between various network functions and services, such as the core network, radio access network, and edge computing resources.

OVS can be used to support network slicing, which is the ability to create virtual networks customized for specific use cases, such as autonomous vehicles, smart cities, or industrial automation. Using the OVS to manage network traffic between these virtual networks, network administrators can ensure that each network slice is isolated and has its own unique network characteristics, such as bandwidth, latency, and reliability.

OVS can also be used to provide network functions virtualization (NFV) capabilities, which allow network functions to be run on commodity hardware instead of proprietary hardware appliances. This can help to reduce costs and increase flexibility in 5G networks. Figures 6.5a,b show how data packets are processed with and without DPDK.

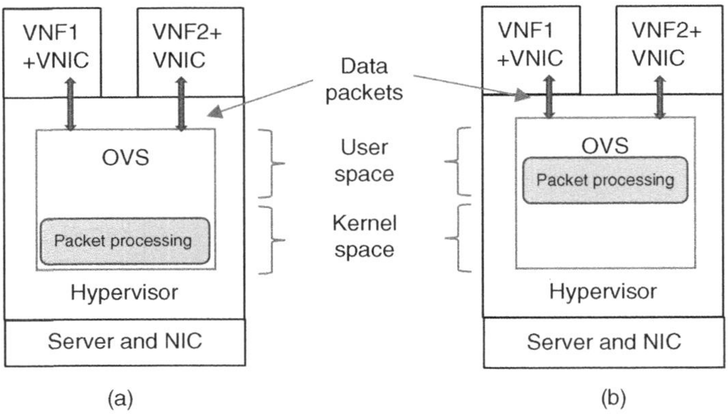

Figure 6.5 Simplified virtual network function with and without DPDK. (a) No DPDK and (b) with DPDK.

As shown in Figure 6.5a, data packet processing is performed in the Kernel space when no DPDK used in OVS. When DPDK is used, the process is performed in the user space, thus improving network performance. Moreover, DPDK can also run on virtual network functions, thus further improving network performance [9, 10].

Another arrangement for DPDK is to use it with a single root input output virtualization (SR-IOV), a technology that allows a single physical network interface to appear as multiple virtual interfaces, each with its own unique identifier and capabilities. In 5G networks, SR-IOV can be used to improve the efficiency of NFV by enabling VNFs to bypass the hypervisor and communicate directly with the physical network interface. This can help to reduce the overhead associated with virtualization and improve overall system performance. It is particularly useful when there are more virtual network functions than the number of physical NICs available. Figure 6.6 illustrates the relationship between fast packet processing with DPDK and virtualization with SR-IOV (Single Root I/O Virtualization).

Although not shown in the illustration above, DPDK can run on the virtual network functions in this arrangement, similar to the OVS scenario. Combining DPDK with SR-IOV can further improve data processing speed and reduce latency in 5G networks. While DPDK facilitates data processing to be done in the user space, SR-IOV enables multiple virtual network functions sharing the same physical NIC. The combination eliminates overheads introduced not only by operating system kernel but also by the hypervisor as well.

Use of DPDK Versus SR-IOV Both DPDK and SR-IOV appear to be good ways to accelerate data processing in network operations. We have discussed how both

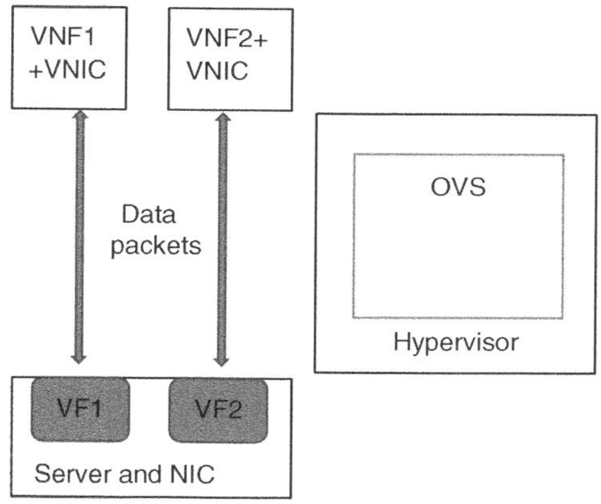

Figure 6.6 Arrangement showing DPDK with SR-IOV.

methods achieve data processing acceleration in the previous sections. In practice, different use scenarios may call for different methods. Intel has conducted a study of the use of DPDK-assisted host and SR-IOV [11] and demonstrated how different methods may produce significant improvement in network performance for different use cases.

In this Intel study, two data flow directions are specified: (i) north–south, in which data is routed to the virtual network function via NIC, processed by the virtual network function(s), and returns to the network and (ii) east–west in which data is processed by a virtual network function and then passed to another (on the same physical layer). Multiple passings of data may take place in a chained form as multiple virtual network functions may be present on the physical layer. Tests are conducted in two configurations: (i) DPDK-assisted device and (ii) SR-IOV with throughput measured in each configuration with data packets of different sizes transmitted through them (i) north–south direction and (ii) east–west direction. Refer to [11] for details of the test setup. Results are shown in Figures 6.7 and 6.8.

In the north–south direction, the performance of SR-IOV is consistently higher across all data packet sizes and all VM configurations over DPDK-assisted devices. The difference in performance is more significant for small packet sizes (e.g., 64 and 128 bytes). In the east–west direction, the performance in DPDK-assisted devices is consistently higher over that of SR-IOV in high data packet sizes (e.g., 1024 and 1518 bytes). DPDK-assisted devices have a significant performance advantage over SR-IOV when the number of VMs is high(e.g., over 4).

Discussion on Test Results The results demonstrate that SR-IOV offers consistent advantages in applications in which data traffic travels in the north–south

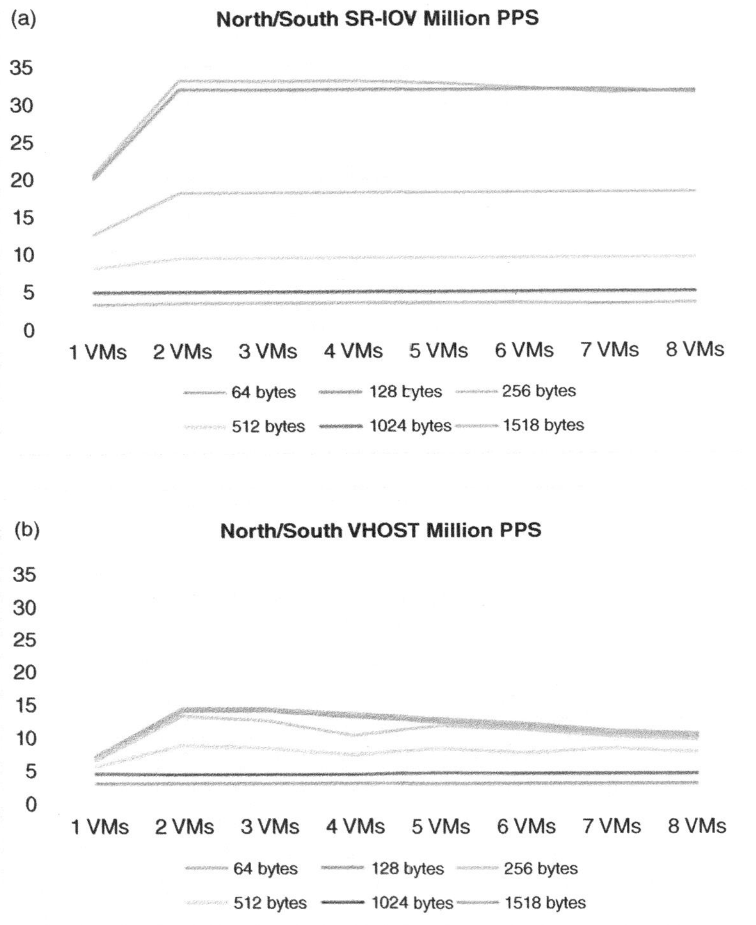

Figure 6.7 Test results in north–south direction [11]/with permission of Intel Corporation. (a) SR-IOV. (b) DPDK-assisted device.

direction. These advantages are more significant when data packets are small in size. However, SR-IOV is not suitable for applications in which data traffic flows in the east–west direction (i.e., from VM to VM). SR-IOV cannot gain any advantage when data is processed or routed within a server where the DPDK-assisted device performs well, especially when the data packet size is high and there is a large number of VMs. The study concludes that the choice between SR-IOV and DPDK depends on data traffic flow direction, and a network designer should consider this point during the design stage.

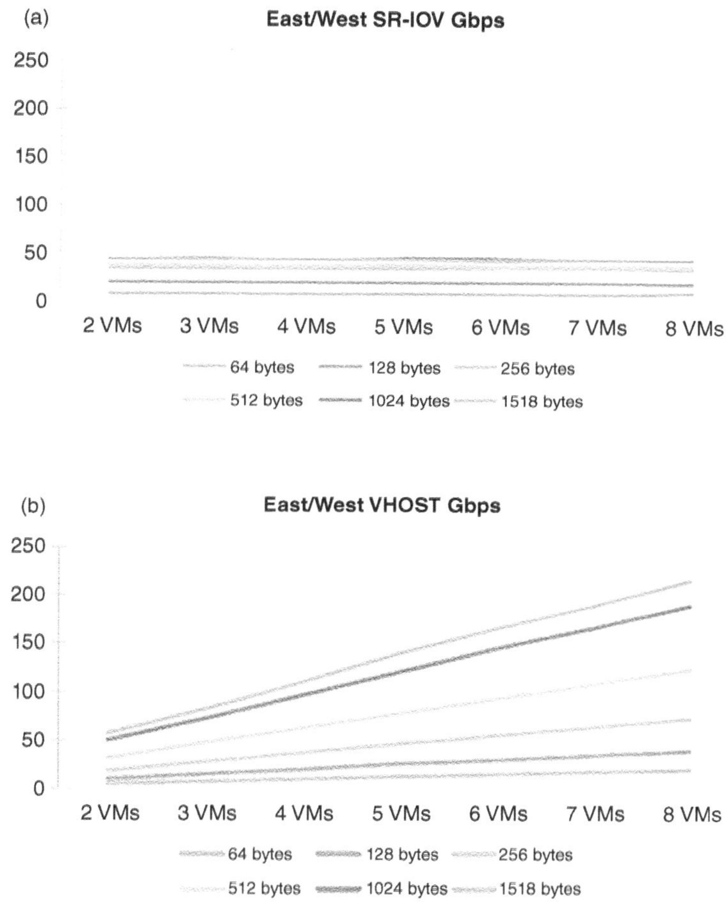

Figure 6.8 Test results in east–west direction [11]/with permission of Intel Corporation. (a) SR-IOV. (b) DPDK-assisted device.

6.3.2 Other Approaches

6.3.2.1 Remote Direct Memory Access (RDMA)

Remote direct memory access (RDMA) is a technology that enables efficient data transfer between two networked devices without involving the CPU of either device. It is widely used in high-performance computing and data center environments [12]. In 5G, RDMA can be used to improve network performance by reducing latency and improving throughput, which is useful for applications such as video streaming, gaming, and virtual reality, where even a small amount of latency can significantly impact user experience.

In 5G, RDMA works by allowing devices to access each other's memory directly without going through the CPU. When a device wants to send data to another device using RDMA, it first requests access to the other device's memory. If the request is granted, it can then write directly to the other device's memory without involving the CPU. This can result in much faster data transfer rates compared to traditional network protocols that require data to be copied multiple times by the CPU in the network stack. Specialized network adapters that support RDMA are used. These adapters are typically equipped with a specialized chip that can access memory directly using direct memory access (DMA) technology. See Figure 6.9 for an illustration.

RDMA can also improve network efficiency by reducing CPU utilization. Traditional network protocols require the CPU to handle all data transfer operations, which can take up a significant amount of CPU resources. By offloading data transfer operations to specialized network adapters, RDMA can free up CPU resources for other tasks.

In 5G, RDMA is typically implemented using the RDMA over Converged Ethernet (RoCE) protocol. RoCE is a standardized protocol that allows for RDMA over Ethernet networks. RoCE can be used with both 5G standalone (SA) and non-standalone (NSA) networks.

RDMA can be used with other 5G technologies, such as network slicing and edge computing, to create more efficient and optimized networks. For example, RDMA can be used to transfer data between edge devices and edge servers, which

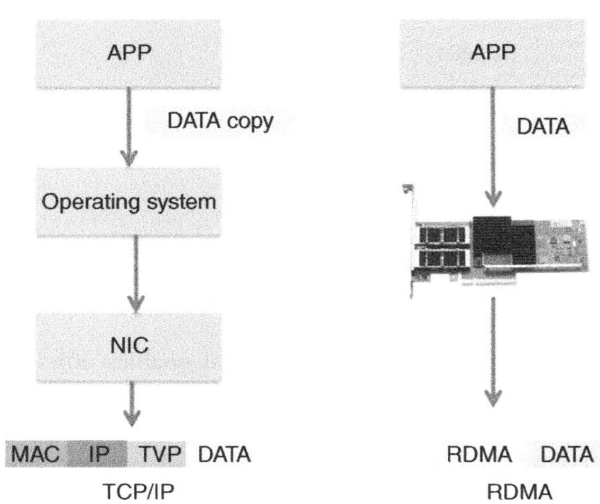

Figure 6.9 Illustration of data of conventional setup (using TCP/IP) and RDMA. Source: https://www.fibermall.com/blog/what-is-rdma.htm.

can reduce latency and improve data transfer rates. Examples of RDMA use in wireless networks include [13, 14].

RDMA is not without its challenges. One of the main challenges of RDMA in 5G is ensuring security. Because RDMA allows for direct access to memory, it can potentially be used to bypass traditional security mechanisms. To address this issue, specialized RDMA-enabled network adapters often include hardware-based security features to prevent unauthorized access.

Software incorporating RDMA offers the following advantages:

Low Latency: RDMA can enable extremely low latency data transfers, which can be critical for certain 5G use cases such as autonomous vehicles or remote surgery. In traditional networks, data transfer involves several steps, including the CPU and operating system processing, which can introduce delays. With RDMA, data can be transferred directly between the memory of two devices, resulting in latency reductions of several orders of magnitude.

High Throughput: RDMA can also enable higher throughput data transfers. With RDMA, data transfer can be much more efficient, resulting in less overhead and more efficient use of network resources. This can enable 5G networks to support more data-intensive applications and services, such as 4 K/8 K video streaming or virtual reality.

Reduced Network Congestion: RDMA can help reduce network congestion by enabling servers to efficiently transfer data without consuming network resources. This can help alleviate network bottlenecks, leading to faster and more reliable data transfers.

Increased Scalability: RDMA can enable 5G networks to scale, supporting more devices and users. By reducing the load on the CPU and operating system of servers, RDMA software can free up resources for other network tasks, allowing the network to handle more traffic and devices.

Improved Reliability: RDMA can improve the reliability of 5G networks by reducing the chances of packet loss and ensuring that data is delivered quickly and efficiently. This can be especially important for real-time applications like video streaming or online gaming.

Efficient Use of Resources: RDMA can help optimize the use of network resources. Traditional network protocols, such as TCP/IP, are designed to be reliable, but they can also consume significant resources, such as CPU time, memory, and network bandwidth. RDMA, on the other hand, is designed to be efficient, enabling data to be transferred with minimal overhead and freeing up network resources for other tasks.

Flexibility: RDMA can be implemented in a variety of ways, including in hardware, firmware, and software. This means it can be customized to meet the specific needs of different 5G applications and services. For example, RDMA can be used with virtualization technologies to enable efficient data transfers between VMs.

Security: RDMA can also help improve the security of 5G networks. By transferring data directly between two devices, RDMA can reduce the number of points in the network where data is vulnerable to interception or tampering.

6.3.2.2 Compute Express Link (CXL)

Although CXL is not directly related to 5G, it is a high-speed interface technology that enables efficient communication between devices such as CPUs, GPUs, and FPGAs [15]. Recent CXL developments have offered much potential [16, 17]. In the context of 5G, CXL can enable the high-speed data processing that is required for advanced 5G applications such as augmented reality, virtual reality, and autonomous vehicles. This is because these applications require high-performance computing resources that can handle large amounts of data quickly and efficiently.

CXL, developed by Intel, provides a low-latency, high-bandwidth interface between processors, memory, and accelerators, enabling these devices to communicate and share data more efficiently. This can result in faster and more efficient data processing, critical for high-performance 5G applications.

CXL uses a cache coherency protocol that ensures that all devices in the system have consistent access to data. It can enable efficient data processing in 5G networks, particularly at the network edge, where data is processed in real time. It can also be used in data centers that support 5G networks, where it can help accelerate data processing and reduce latency.

CXL Protocols CXL protocols define the communication interface between the host CPU and the CXL devices, allowing them to exchange data with minimal latency and high bandwidth. The CXL protocols include:

CXL.io: This protocol defines the physical and electrical interface between the host CPU and the CXL device. It includes the electrical signaling, clocking, and data encoding schemes.

CXL.cache: This protocol defines the cache coherence between the host CPU and the CXL device. It allows the CXL device to access the host CPU's memory as if it were its own, reducing latency and increasing bandwidth.

CXL.mem: This protocol defines the memory coherence between the host CPU and the CXL device. It enables the CXL device to directly access the host CPU's memory without the need for copying data between the two devices.

CXL protocols are illustrated in Figure 6.10.

The CXL protocol is designed to improve the performance of data-intensive applications by allowing for high-speed communication between processors,

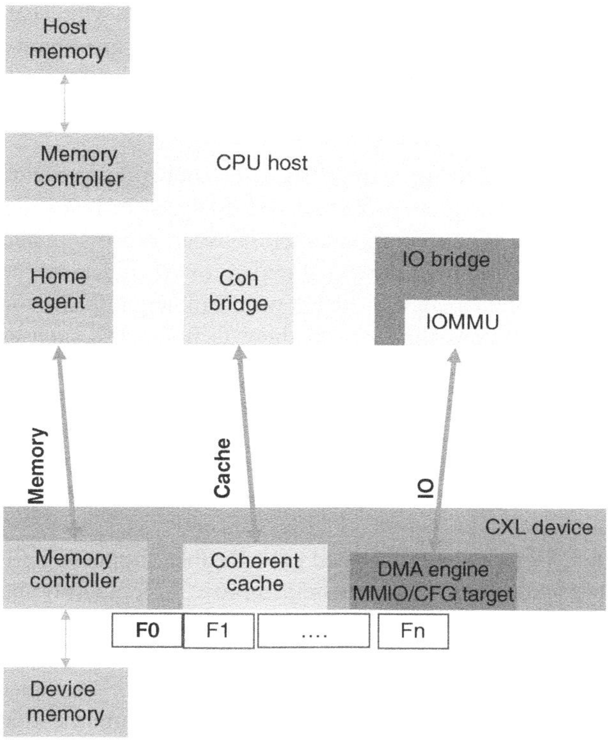

Figure 6.10 CXL protocols. Source: Adapted from CXL Consortium.

accelerators, and memory. While CXL is not directly related to 5G networks, it can indirectly improve the 5G network performance by enabling faster processing and analysis of the large amounts of data that 5G networks generate.

Here are a few ways that the CXL protocol can improve 5G network performance:

Accelerated Data Processing: CXL allows faster communication between processors and accelerators, which can help accelerate data processing and reduce latency. This can be especially useful for processing large amounts of data generated by 5G networks, such as video streams, sensor data, and other types of IoT data.

Increased Memory Bandwidth: CXL can increase the memory bandwidth available to processors and accelerators, which can help improve their performance when working with large datasets. This can be particularly beneficial for applications that require real-time data processing and analysis, such as autonomous vehicles and industrial automation.

Improved Energy Efficiency: CXL is designed to be energy efficient, which can help reduce the power consumption of data centers and other high-performance computing environments. This can be important for 5G networks, which require a significant amount of power to operate and can benefit from energy-efficient technologies.

Improved AI and Machine Learning Capabilities: 5G networks generate a vast amount of data that needs to be processed and analyzed in real time. The CXL protocol can enable faster communication between processors and accelerators, which can help improve the performance of AI and machine learning algorithms used for data analysis. This, in turn, can help improve the accuracy and speed of applications that rely on these technologies, such as autonomous vehicles, smart cities, and healthcare systems.

Enhanced Virtualization: CXL can also improve the performance of virtualized environments by enabling faster communication between VMs and the underlying hardware. This can help reduce the overhead associated with virtualization and improve the performance of applications running on virtualized infrastructure.

Greater Flexibility: The CXL protocol is designed to be flexible, supporting different topologies and configurations, which can help improve the scalability and versatility of 5G networks. This can be especially important as 5G networks continue to evolve and support a wider range of applications and devices.

Reduced Complexity: CXL can help reduce the complexity of high-performance computing environments by providing a common interface for communication between processors, accelerators, and memory. This can simplify the design and management of these environments, which can improve their reliability and reduce their operational costs.

6.3.2.3 Data Processing Unit (DPU)

A DPU in 5G is a specialized hardware component designed to accelerate data processing tasks in the network. It is typically used to perform high-speed data processing functions such as packet inspection, encryption/decryption, and protocol conversion [18, 19].

With 5G networks, the amount of data that needs to be processed is significantly higher than in the earlier generations of wireless technologies. This is due to the large number of connected devices, the increased use of high-quality video streaming, and the growing popularity of cloud-based applications. To handle the massive amounts of data in 5G networks, DPUs are used at various points within the network. For example, in the radio access network, DPUs are used to process data from mobile devices and transmit it to the core network. In the core network,

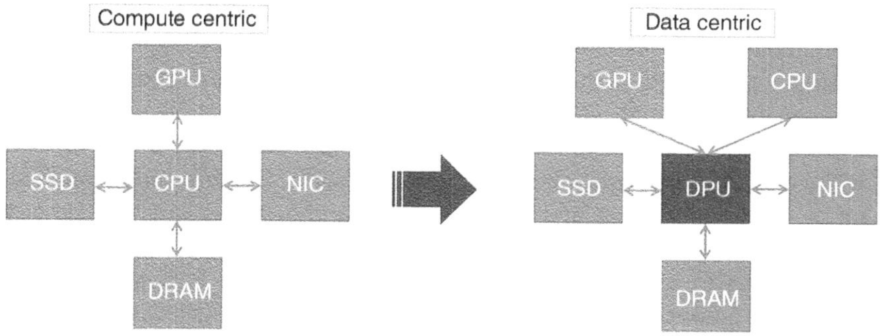

Figure 6.11 Example use of DPU [23].

DPUs are used to handle data traffic, manage network resources, and route data to its destination.

DPUs are typically used in conjunction with NICs to handle the high data rates and low latency requirements of 5G applications. Examples include [20–22]. See Figure 6.11 for an illustration. DPUs are designed to work in parallel with the central processing unit (CPU) of a server. One of the main benefits of using DPUs in 5G networks is that they can offload processing tasks from the CPU. DPUs are designed to handle specific tasks, such as packet inspection or encryption/decryption, much more efficiently than a general-purpose CPU. By offloading these tasks to a DPU, the CPU is freed up to handle other tasks, which can improve overall system performance and reduce latency.

The following paragraphs highlight the advantages offered by DPUs in 5G networks.

A key advantage of DPUs in 5G networks is that they can be used to implement NFV. NFV allows network functions such as firewalls, load balancers, and intrusion detection systems to be implemented in software running on VMs rather than as dedicated hardware appliances. This approach can be more cost-effective and flexible than traditional hardware-based approaches, but it can also require a significant amount of processing power. By using DPUs to accelerate the processing of these virtual network functions, the overall system performance can be improved while maintaining the flexibility and cost-effectiveness of NFV.

DPUs can also be used to implement SDN in 5G networks. SDN allows network administrators to manage network traffic more dynamically by separating the CP from the DP. By offloading some of the processing tasks to a DPU, SDN controllers can make faster and more informed decisions about how to route the network traffic, which can improve overall system performance and reliability.

Edge computing is a key trend in 5G networks, as it enables faster data processing and reduces network latency. DPUs deployed at the network edge can process data in real time, without the need to send it back to the core network. This can help to enable new applications such as smart cities, autonomous vehicles, and augmented reality, which require low latency and high bandwidth.

DPUs can also help improve security in 5G networks. By performing functions such as packet inspection and encryption/decryption, DPUs can help protect against cyber-attacks and ensure that data is transmitted securely.

5G networks are expected to support a large number of IoT devices, which will generate vast amounts of data. DPUs can handle this data by aggregating, analyzing, and acting on it in real time. This can help to enable new IoT applications, such as smart agriculture, smart manufacturing, and smart homes.

Finally, another advantage of using DPUs in 5G networks is that they can help reduce power consumption. DPUs are designed to be highly efficient, and they can perform specific functions using much less power than a general-purpose CPU. This can be extremely important in 5G networks, where there is a growing need to reduce energy consumption and increase sustainability.

References

1 Dutta, A. and Hammad, E. (2020). 5G security challenges and opportunities: a system approach. *2020 IEEE 3rd 5G World Forum (5GWF)*, 10 September–10 October 2020, Virtual Conference Bangalore India, pp. 109–114.

2 3GPP (2017). TS23.501 System Architecture for the 5G System, Stage 2.

3 The Data Plane Development Kit Home Page. http://www.dpdk.org/ (accessed on 1 September 2024).

4 The Linux Foundation Home Page. http://www.Linuxfoundation.org/ (accessed on 1 September 2024).

5 Pak, J. and Park, K. (2018). A high-performance implementation of an IoT system using DPDK. *Applied Sciences* 8 (4): 550.

6 Zhang, H., Chen, Z., and Yuan, Y. (2021). High-Performance UPF Design Based on DPDK. *2021 IEEE 21st International Conference on Communication Technology (ICCT)*, 13–16 October 2021, Tianjin China, pp. 349–354.

7 Askari, L., Majidzadeh, P., Ayoub, O., and Tornatore, M. (2020). Exploiting DPDK in containerized environment with unsupported hardware. *2020 IEEE Conference on Network Function Virtualization and Software Defined Networks (NFV-SDN)*, 10–12 November 2020, Virtual Conference Madrid/Leganés Spain, pp. 7–12.

8 The Open Virtual Switch Home Page. http://www.openvswitch.org/ (accessed on 1 September 2024).

9 Garbugli, A., Rosa, L., Foschini, L. et al. (2022). A Framework for TSN-enabled virtual environments for ultra-low latency 5G scenarios. *ICC 2022 – IEEE International Conference on Communications*, 16–20 May 2022, Seoul South Korea, pp. 5023–5028.

10 Chun, B., Ha, J., Oh, S. et al. (2019). Kubernetes enhancement for 5G NFV infrastructure. *2019 International Conference on Information and Communication Technology Convergence (ICTC)*, 16–18 October 2019, Jeju Island South Korea, pp. 1327–1329.

11 Kutch, P. and Johnson, B. (February 2017). SR-IOV for NFV solutions, practical considerations and thoughts. Technical Brief, *Technical Brief, Intel Corporation Networking Division.* https://www.intel.com/content/dam/www/public/us/en/documents/technology-briefs/sr-iov-nfv-tech-brief.pdf (accessed on 1 September 2024).

12 Barghash, A., Hammad, L., and Gharaibeh, A. (2022). Traditional vs. modern data paths: a comprehensive survey. *Computers* 11 (9): 132.

13 Varga, P., Peto, J., Franko, A. et al. (2020). 5G support for industrial IoT applications – challenges, solutions, and research gaps. *Sensors* 20 (3): 828.

14 Li, T., Xu, K., Huang, H. et al. (2022). WIP: when RDMA meets wireless. *2022 IEEE 23rd International Symposium on a World of Wireless, Mobile and Multimedia Networks (WoWMoM)*, 14–17 June 2022, Belfast United Kingdom, pp. 177–180.

15 Compute Express Link (CXL) Specifications Home Page. CXL Consortium Incorporated. https://www.computeexpresslink.org/ (accessed on 1 September 2024).

16 Das Sharma, D. (2021). Keynote 1: Compute Express Link (CXL) changing the game for cloud computing. *2021 IEEE Symposium on High-Performance Interconnects (HOTI)*, 18–20 August 2021, Santa Clara California USA, xii–xii.

17 Coughlin, T. (2022). Digital storage and memory. *Computer* 55 (1): 20–29.

18 Deierling, K. (2020). What is a DPU? *NIVIDA Blog*, 20 May 2020. https://blogs.nvidia.com/blog/2020/05/20/whats-a-dpu-data-processing-unit (accessed on 1 September 2024).

19 Deierling, K. (2021). Achieving a Cloud-Scale Architecture with DPUs. *NIVIDA Technical Blog*, 18 June 2021. https://developer.nvidia.com/blog/achieving-a-cloud-scale-architecture-with-dpus (accessed on 1 September 2024).

20 Barsellotti, L., Alhamed, F., Vegas Olmos, J.J. et al. (2022). Introducing Data Processing Units (DPU) at the edge [Invited]. *2022 International Conference on Computer Communications and Networks (ICCCN)*, 25–28 July 2022, Honolulu Hawaii USA, pp. 1–6.

21 Pismenny, B., Olmos, J.V., Zack, Y., et al. (2021). Securitization of cloud, edge and IoT communications through hardware accelerations/offloadings, *2021*

IEEE International Mediterranean Conference on Communications and Networking (MeditCom), 7–10 September 2021, Athens Greece, pp. 61–64.

22 Ramachandran, P., Ranganath, S., Bhandaru, M., and Tibrewala, S. (2021). A survey of AI enabled edge computing for future networks. *2021 IEEE 4th 5G World Forum (5GWF)*, 13–15 October 2021, Montreal QC Canada, pp. 459–463.

23 Yoshida, H. (2020). Moving to data centric data centers and DPUs. *Hitachi Blog*, 6 October 2020. https://community.hitachivantara.com/blogs/hubert-yoshida/2020/10/06/moving-to-data-centric-data-centers-and-dpus (accessed on 1 September 2024).

Part IV

Applications

7

Killer Applications

The concept of a "killer app" refers to a software application or technology that is so innovative, useful, or popular that it becomes the driving force behind the adoption of a particular platform, device, or technology. A killer app is essentially an application or feature so compelling that people will buy a new device or adopt a new technology just to use it.

Metaverse has the potential to create an entirely new world of possibilities, transforming how we interact with each other, consume media and entertainment, learn, work, and do business. As more people begin to explore the metaverse and develop new applications and use cases, its impact could be profound and far-reaching, making it a potential future technology killer application. In the following sections, we will look at what metaverse is, the technologies behind it, and its potential applications.

7.1 Metaverse and Its Trends

The term "metaverse" was first introduced by science fiction author Neal Stephenson in his 1992 novel "Snow Crash" [1]. In the book, the metaverse is a virtual reality (VR) universe where people can interact with each other and engage in various activities. Since then, the term has been used to refer to a wide range of virtual and augmented reality (AR) experiences, and there is growing interest in the potential of the metaverse to become the next major platform for online interaction and commerce.

The metaverse is a complex concept that involves multiple layers of technology, platforms, and experiences [2–5]. It can be thought of as a fully realized virtual world that users can interact with and explore [6]. The metaverse is becoming a

Edge Computing Acceleration: From 5G to 6G and Beyond, First Edition.
Patrick Hung, Hongwei Kan, and Greg Knopf.

reality thanks to technological advancements, such as VR, AR, blockchain, and artificial intelligence. The metaverse is essentially a collective virtually shared space that combines the physical and digital worlds. It is a fully immersive environment where people can engage in various activities, interact with others, and even conduct business [7].

One of the key benefits of the metaverse is the ability to create new experiences that are not possible in the physical world. People can explore virtual worlds that are based on science fiction or fantasy, and they can interact with fictional characters in a more immersive way [8].

Another benefit of the metaverse is the potential for new economic opportunities. As the metaverse grows, there will be opportunities for new businesses and entrepreneurs to create virtual goods and services. This includes everything from virtual real estate to virtual clothing to virtual education and training [9–14].

Here are some of the key components that make up the metaverse:

Virtual Environments: The metaverse comprises various virtual environments, such as three-dimensional (3D) worlds, gaming environments, social platforms, and virtual marketplaces. These environments can be accessed through various devices, including computers, mobile phones, and VR headsets.

Avatars: In the metaverse, users can create avatars representing them in the virtual world. Avatars can be customized with different clothing, hairstyles, and other accessories, allowing users to express their individuality [9, 15].

Interactions: The metaverse is designed to be a social environment, and users can interact with each other in various ways. This includes chatting, messaging, voice and video calls, and participating in multiplayer games and events [15].

Transactions: One of the key features of the metaverse is the ability to conduct transactions using virtual currencies. This includes buying and selling virtual items, such as clothing, accessories, and even virtual real estate [7, 9].

Blockchain Technology: The metaverse is built on blockchain technology, which provides a secure and decentralized infrastructure for conducting transactions and managing digital assets. This allows users to have full control over their virtual assets and ensures they cannot be duplicated or counterfeited [16].

However, there are also potential risks and challenges associated with the metaverse. One of the concerns is privacy and security [17]. As people engage in virtual activities and transactions, there is a risk that their personal information could be compromised. Additionally, there is a risk that the metaverse could be used for illegal activities, such as money laundering or terrorism.

Overall, the metaverse is a complex ecosystem that involves multiple technologies and platforms working together to create a seamless virtual world that users can explore and engage with.

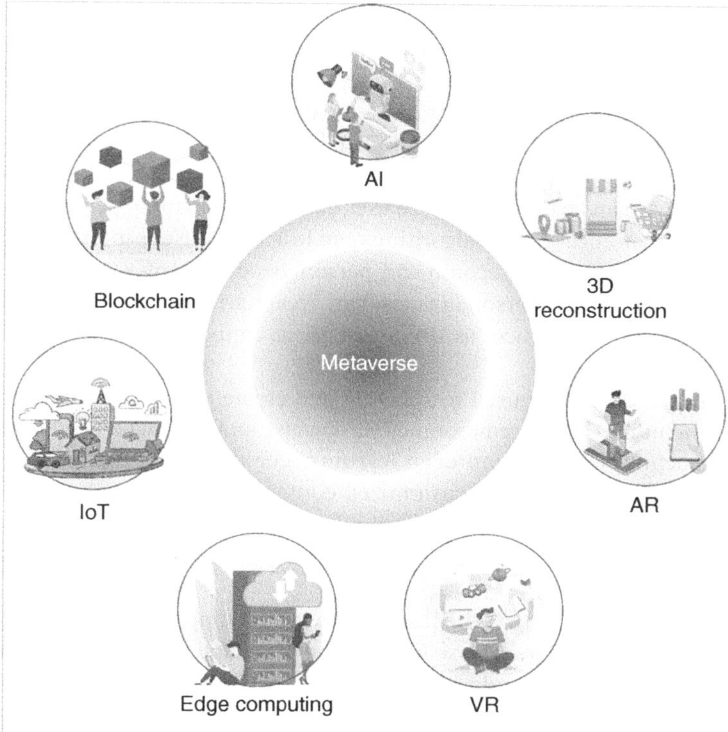

Figure 7.1 Technologies supporting metaverse. Source: [2] Al-Ghaili, A.M et al., 2022/IEEE/CC BY 4.0.

7.2 Technologies Behind Metaverse

Metaverse refers to a hypothetical future iteration of the Internet that is immersive, 3D, and interactive. While the metaverse concept is still largely speculative, several technologies are already laying the groundwork for its development. Here are some of the key technologies behind the metaverse (Figure 7.1).

7.2.1 Artificial Intelligence

AI has the potential to significantly enhance the metaverse experience by making it more personalized, engaging, and secure. As the metaverse continues to evolve and grow, we can expect AI to play an increasingly important role in shaping the future of this exciting new virtual world [18, 19]. AI can help metaverse in several ways, as listed below:

Intelligent Avatars: AI can help make avatars in the metaverse more lifelike and expressive, allowing for more realistic interactions and engagement with other users [20]. AI algorithms can generate natural facial expressions and body movements, making avatars more expressive and engaging. Additionally, AI can help customize avatars to reflect users' individual styles and preferences, further enhancing the user experience.

Personalization: AI can be used to analyze user data and behavior to personalize the metaverse experience [21]. AI algorithms can learn from user's interactions, preferences, and activities to recommend personalized content, experiences, and virtual environments. Personalization can create a more immersive and engaging metaverse experience, as users feel more connected to the virtual world they are exploring.

Content Creation: AI can be used to generate and create new content for the metaverse, including virtual environments, objects, and characters [20]. By leveraging machine learning and deep learning techniques, AI can analyze existing content and create new, customized content at scale. This can help reduce the amount of time and resources required for manual content creation while also increasing the diversity and quality of content available in the metaverse.

Natural Language Processing (NLP): AI can be used to improve communication and interaction in the metaverse by enabling more natural and intelligent conversations between users and virtual assistants [20]. NLP algorithms can be used to understand and interpret natural language inputs from users, making it easier for them to communicate their needs and preferences. Additionally, AI-powered chatbots can assist users with various tasks, such as finding content or providing customer support.

Predictive Analytics: AI can be used to analyze user data and behavior to make predictions about future actions and preferences. By analyzing user data, AI can identify patterns and trends that can be used to personalize recommendations, predict user behavior, and enhance the overall user experience. This can help keep users engaged and interested in the metaverse, as they are provided with content and experiences that are tailored to their interests and preferences [20, 22].

Fraud Detection: AI can be used to detect and prevent fraud in the metaverse [23]. For instance, AI algorithms can be used to analyze user behavior and identify patterns that are indicative of fraudulent activity, such as abnormal login attempts or suspicious transactions. By detecting and preventing fraud, AI can help maintain a safe and secure environment for all users in the metaverse.

In addition to the above, there are two areas of non-player character (NPC) and senses where AI can assist metaverse. They are discussed in the following two subsections.

7.2.1.1 AI-Based Non-player Character

An NPC (Non-player Character) is a character in a video game or other virtual environment that is controlled by the game's AI rather than by a human player [24, 25]. NPCs are often used to populate game worlds with characters that players can interact with, such as merchants, quest givers, and enemies.

In most cases, NPCs in video games are programmed using AI techniques such as decision trees, pathfinding algorithms, and behavior trees. These techniques enable NPCs to perform a variety of actions and behaviors in response to player actions or game events. For instance, an NPC merchant might use AI to determine what items to sell, how much to charge for them, and how to respond to player requests. An NPC enemy might use AI to determine when to attack, when to retreat, and how to evade the player's attacks.

The following are brief descriptions of the key AI techniques used in NPCs [20, 26].

Decision trees are a popular AI technique used to model decision-making processes in NPCs. Decision trees can be used to simulate NPC behavior based on a set of rules and conditions. An NPC might decide to attack the player if their health drops below a certain level or to flee if outnumbered.

Pathfinding algorithms are another AI technique used in NPCs to help them navigate through complex game environments. These algorithms can determine the shortest path from one point to another, avoiding obstacles and other NPCs along the way.

Behavior trees are a more complex AI technique that can be used to model complex decision-making and goal-oriented behavior in NPCs. They can be used to simulate NPC behavior based on a hierarchy of goals and subgoals. An NPC might have the overall goal of defeating the player, with subgoals such as attacking, defending, and retreating based on the current situation.

Machine learning techniques can also be used to program NPC behavior. For example, reinforcement learning can be used to train NPCs to learn from their experiences in the game and improve their behavior over time. This can create more adaptive and challenging NPCs that can provide a more engaging gaming experience for players.

7.2.1.2 Sensory Capabilities with AI

AI can help provide sensory capabilities in applications by analyzing data from sensors and other sources in order to provide useful feedback to users [26–28]. By using machine learning techniques to make sense of the environment, AI can help create more intelligent and responsive applications that provide a better user experience. Here are details on how AI can be used to produce tactile, smell, and taste senses:

Tactile: Haptic feedback systems use various types of actuators, such as motors or piezoelectric materials, to create vibrations or forces that can simulate the sense of touch [26]. AI can be used to optimize the design of these systems by analyzing the data from sensors that detect the user's movements, pressure, or contact with the device. By using machine learning algorithms to process this data, the system can learn to adjust the haptic feedback in real time, based on the user's preferences and the task at hand. For example, a haptic feedback system for a video game controller could use AI to generate different types of feedback for different types of interactions, such as a soft vibration for a gentle touch and a stronger vibration for a hit or impact.

Smell: Artificial olfactory systems use chemical sensors that detect the presence of specific odor molecules in the air. These sensors can be arranged in an array and used to detect different combinations of molecules, which can be analyzed and interpreted by AI algorithms. By training these algorithms on large datasets of chemical signatures, the system can learn to identify different odors and generate corresponding olfactory feedback [27]. An artificial olfactory system could be used in a museum exhibit to simulate the smells of different time periods or environments.

Taste: Artificial taste systems use sensors that detect the presence of specific taste molecules, such as sweet, salty, bitter, or sour compounds. These sensors can also be arranged in an array and used to detect different combinations of molecules, which can be analyzed and interpreted by AI algorithms. By training these algorithms on large datasets of taste profiles, the system can learn to identify different flavors and generate corresponding taste feedback [28]. An artificial taste system could be used in a food production facility to detect and control the flavor of products or in a healthcare setting to help individuals with taste disorders.

In all these applications, AI plays a critical role in optimizing the design of the sensory feedback systems and enabling more precise and nuanced interactions with technology. As these technologies continue to advance, we can expect to see new and exciting applications emerge that enhance our sensory experiences and improve our quality of life.

7.2.2 Blockchain

Blockchain technology can provide various benefits for the metaverse, including creating more secure and transparent in-game economies, enabling greater user control over digital assets, supporting interoperability between different virtual worlds and platforms, and promoting more democratic and transparent community governance [16, 29]. These benefits can help create a more immersive and engaging virtual world where users can explore, create, and interact with each other in new and exciting ways. The following is a look at how blockchain can support the metaverse.

Decentralized Ownership: Blockchain technology can allow for decentralized ownership of virtual assets, meaning that users can have full control over their digital property without needing a centralized authority or platform [16, 30]. This can be achieved using non-fungible tokens (NFTs), which are unique digital assets that can be bought, sold, and traded on blockchain networks. In the metaverse, NFTs can represent virtual land, buildings, items, and other in-game assets, allowing users to have full ownership and control over their virtual possessions.

Immutable Records: Blockchain technology can create a tamper-proof and transparent record of all transactions within the metaverse [30]. This can help prevent fraud and protect users from theft or loss of their virtual assets. By using blockchain-based ledgers, transactions can be recorded in a decentralized and secure manner, creating a permanent record that cannot be altered or deleted.

In-Game Economies: Blockchain technology can enable the creation of more robust and secure in-game economies within the metaverse [29]. By using blockchain-based currencies, users can have greater confidence in the security and reliability of these virtual currencies. Additionally, blockchain-based smart contracts can be used to automate complex in-game transactions, reducing the risk of human error and fraud. Smart contracts can also enforce in-game rules and regulations, creating a more transparent and fair virtual world.

Interoperability: Blockchain technology can facilitate interoperability between different virtual worlds and platforms within the metaverse [20]. By using blockchain-based standards and protocols, virtual environments can communicate and share data with each other, creating a more seamless and integrated experience for users. This can also enable users to move their digital assets between different virtual worlds and platforms, creating a more interconnected and dynamic metaverse.

Community Governance: Blockchain technology can enable more democratic and transparent community governance within the metaverse [29]. Using blockchain-based voting mechanisms, users can participate in the decision-making processes that govern the virtual world. This can create a more inclusive and participatory environment where users have a greater say in the development and management of the metaverse.

7.2.2.1 Power Consumption

Blockchain has the potential to help reduce power consumption in 5G by enabling more efficient management of network resources. By having a reduced 5G power consumption, the metaverse power efficiency will also improve. Here are a few ways blockchain can be used to improve the power efficiency in 5G and in the Metaverse in turn:

Decentralized Network Management: One of the key features of blockchain is its ability to support decentralized network management [30]. This means that instead of relying on a centralized authority to manage network resources, blockchain can enable a decentralized network of nodes to manage resources more efficiently. For example, blockchain could be used to manage the allocation of network resources, such as bandwidth, among different users and applications. By using a decentralized approach, blockchain could help ensure that resources are allocated more efficiently and that power consumption is minimized.

Smart Contracts: Smart contracts are self-executing and enforcing contracts with the terms of the agreement between buyer and seller being directly written into lines of code [29]. Smart contracts can automate certain processes, such as billing and payment processing. By automating these processes, blockchain can help reduce the energy required to manage these tasks manually, thus reducing the overall power consumption.

Energy Trading: Another potential use case for blockchain is energy trading [31]. Blockchain could be used to enable energy trading among different users and applications. For example, a user could sell excess energy generated by their solar panels to another user who needs more energy. By using blockchain to manage energy trading, the network could help ensure that energy is used more efficiently, which could help reduce the overall power consumption.

7.2.3 AR and VR

AR and VR technologies can help enhance the metaverse by creating a more immersive and interactive experience for players, enabling real-time interaction between players from different parts of the world, enabling mixed-reality experiences, adding gamification elements, and enhancing personalization [27]. These technologies can make the metaverse more engaging and exciting and can help create a more seamless and integrated gaming experience for players. The following are some examples:

Enhanced Immersion: AR and VR can make the metaverse more immersive by creating a more realistic and engaging environment [8, 32]. For example, players can use VR headsets to fully immerse themselves in the virtual world and interact with it in a more natural and intuitive way.

Real-Time Interaction: AR and VR can enable real-time interaction between players in the metaverse, even if they are in different parts of the world [33]. This can enhance the social aspect of the metaverse and create a more engaging and interactive experience for players.

Mixed Reality Experiences: AR and VR can enable mixed reality experiences in the metaverse, where real-world and virtual-world elements are blended seamlessly [33]. Players can use AR technology to overlay virtual elements onto their real-world surroundings, creating a more immersive and interactive experience.

Gamification: AR and VR can add gamification elements to the metaverse, making it more fun and engaging for players [34]. For instance, players can use AR technology to complete quests or challenges in the real world or use VR technology to fully immerse themselves in virtual games within the metaverse.

Personalization: AR and VR can enhance the personalization of the metaverse by allowing players to create and customize their own avatars, environments, and experiences [12, 35]. Players can use VR technology to create custom avatars that fully represent their personalities and preferences or use AR technology to create customized environments that reflect their interests.

7.2.4 Internet of Things

IoT has the potential to significantly enhance the metaverse experience by providing real-world data, enabling physical interactions to control virtual objects, providing personalized recommendations, and creating more immersive and interactive experiences [36]. We can expect to see more integration between IoT and metaverse technologies in the future. The key areas in which IoT can support metaverse are as follows:

Sensor Data: IoT sensors can collect real-world data that can be used to create more immersive and interactive metaverse experiences [37]. Sensors embedded in physical spaces can collect data on temperature, humidity, and other environmental factors, which can be used to create more realistic virtual environments. Similarly, sensors embedded in physical objects can collect data on how those objects are used, which can be used to create more realistic virtual objects.

Wearables: Wearable IoT devices can provide users with real-time feedback on their physical movements and actions, which can be used to control virtual avatars and objects in the metaverse [38]. A user wearing a motion sensor could use their physical movements to control the movement of their virtual avatar in the metaverse. Wearables can also provide feedback on biometric data such as heart rate and stress levels, which can be used to create more engaging and personalized metaverse experiences.

Smart Objects: IoT-enabled smart objects can be integrated into the metaverse to provide users with more interactive and personalized experiences [39]. A user could use a smartwatch to control the lighting and temperature in their virtual home or use a smart speaker to interact with virtual assistants or other characters in the metaverse. Smart objects can also provide users with personalized recommendations and other forms of assistance based on their behaviors and preferences.

AR: IoT-enabled AR devices can be used to overlay virtual objects and information onto the physical world, creating more immersive and interactive metaverse experiences [40]. A user wearing an AR headset could see virtual objects and characters overlaid onto their physical environment, creating a blended reality experience. IoT sensors can be used to collect data on the user's physical environment, which can be used to create more realistic and accurate virtual overlays.

7.3 Applications of Metaverse

Metaverse is a concept that describes a fully realized virtual world that spans multiple interconnected virtual environments. As such, it can support a wide range of applications and use cases. Here are some of the main applications of the metaverse.

7.3.1 Gaming

The metaverse can enhance the gaming experience by creating a more immersive, social, and dynamic environment [20]. It can provide new opportunities for players to engage with their favorite games. It can encourage creativity and innovation within the gaming community. Here are some examples:

Enhanced Immersion: The metaverse creates a fully realized virtual world that can be explored in a variety of ways. This level of immersion can make gaming more engaging and memorable, leading to increased player satisfaction and retention [41]. For example, in a virtual world, players can fully interact with their surroundings and explore environments that are not possible in the real world. This can make games more enjoyable and immersive, thus increasing player loyalty and attracting new players.

Social Interaction: The metaverse can create a more social and collaborative gaming experience by allowing players to interact with each other in various ways [41, 42]. Players can participate in virtual events, join virtual communities, and play games with other players from around the world. This can foster a sense of community and make gaming more enjoyable for players. In addition, social interaction can encourage teamwork and cooperation, which can improve players' performance in games.

Virtual Economy: The metaverse can create a virtual economy where players can buy, sell, and trade virtual goods and services using virtual currency [42, 43]. This can create new opportunities for players to earn rewards, build their gaming assets, and even earn a living by playing games. Players can buy and sell virtual items such as weapons, armor, and other game-related items. This can create a new revenue stream for game developers and publishers. It can also provide players new ways to engage with their favorite games.

Cross-Platform Play: The metaverse can make it easier for players to connect and play games across different platforms and devices [44]. This can increase the player base for games and make it easier for players to find opponents and teammates. A player using a PC can play a game with someone using a console or mobile device. This can create a more seamless and integrated gaming experience that allows players to play their favorite games with friends and family, regardless of their platform.

Creative Opportunities: The metaverse allows players to create and share their own content, such as custom avatars, game mods, and levels [42]. This can encourage creativity and innovation within the gaming community and lead to the creation of new and exciting gaming experiences. Players can create their own games within the virtual world, which can be shared with other players. This can foster a sense of community and engagement within the gaming community and allow players to create their own unique gaming experiences.

7.3.2 Education

Metaverse has the potential to transform education by providing engaging and interactive learning experiences, accessible education, collaboration and communication opportunities, personalized learning experiences, and the use of cutting-edge technologies like AR and VR [45, 46].

Enhanced Learning Experiences: The metaverse can create virtual worlds that provide students with engaging and interactive learning experiences [47]. These virtual environments can allow students to explore and experiment with concepts and ideas in ways that are not possible in traditional classrooms. For example, students can visit virtual museums or archaeological sites, manipulate objects, and interact with historical figures or experts in the field. This type of immersive learning can make education more engaging, enjoyable, and memorable.

Accessible Education: The metaverse can provide education to people who might not have access to traditional education due to geographic, financial, or other barriers. Online learning can be more affordable and accessible than traditional education, as it does not require physical classrooms, textbooks, or other expensive resources [47]. This can open up opportunities for people in remote or rural areas, as well as those who cannot afford traditional education.

Collaboration and Communication: The metaverse can enable collaboration and communication among students and teachers from around the world [48, 49]. This can provide opportunities for students to learn from people with different backgrounds, cultures, and perspectives. The metaverse can also facilitate teamwork, group projects, and other forms of collaboration that can enhance critical thinking, problem-solving, and communication skills.

Personalized Learning: The metaverse can enable personalized learning experiences based on students' individual needs, interests, and learning styles [50]. The metaverse can use data analytics and artificial intelligence to identify areas where students need extra help, provide targeted feedback, and create personalized learning paths. This can help students learn at their own pace and in ways that work best for them.

AR and VR: The metaverse can enable the use of AR and VR technologies in education [46, 51]. These technologies can help students visualize complex concepts, develop spatial reasoning skills, and explore topics in a more engaging and interactive way. For example, students can use AR to visualize the inner workings of a machine or VR to explore a historical event in 3D.

7.3.3 Commerce

Metaverse has the potential to provide businesses with new opportunities for revenue generation, marketing, and customer engagement [52, 53]. By creating virtual storefronts and experiences, businesses can reach new audiences, increase brand awareness, and create personalized and engaging shopping experiences. As the metaverse continues to develop, we can expect to see new and innovative ways for businesses to leverage this new platform to create value for their customers and stakeholders.

New Revenue Streams: The metaverse can provide businesses with new opportunities to generate revenue by creating virtual products and services [54]. Businesses can sell virtual goods such as clothing, accessories, furniture, and more. Virtual goods can be customized and personalized, providing users with a unique shopping experience. This can create new revenue streams for businesses that are not tied to physical products or traditional e-commerce channels.

Increased Brand Awareness: The metaverse can also provide businesses with a new platform for marketing and advertising [52]. By creating virtual storefronts, businesses can showcase their products and services in a highly engaging and interactive environment. Virtual experiences can provide an immersive way for businesses to showcase their brand, increase awareness, and generate buzz.

Enhanced Customer Engagement: The metaverse can provide businesses with new ways to engage with customers [27]. Virtual experiences and events can allow businesses to interact with distant customers in real time, providing them with a unique and memorable experience. This can help businesses build stronger relationships with their customers, create brand loyalty, and generate positive word-of-mouth.

Accessible Commerce: The metaverse can provide commerce opportunities to people needing access to traditional brick-and-mortar stores or e-commerce websites. The metaverse can be accessed from anywhere in the world, providing opportunities for businesses to reach new markets and expand their customer base [55, 56]. This can help businesses increase their revenue and customer base, regardless of their physical location.

Innovative Business Models: The metaverse can enable new and innovative business models that are not possible in traditional commerce [54, 57]. Businesses can create virtual marketplaces where users can buy and sell virtual goods and services. Businesses can also offer virtual experiences and services, such as virtual concerts, workshops, and consultations [53]. These new business models can create unique value propositions and revenue streams for businesses that are not possible in traditional commerce.

7.3.4 Social Networking

Metaverse has the potential to transform social networking by providing users with a more engaging, immersive, and personalized social experience [33]. Moving forward, we expect to see new and innovative ways for users to connect and engage with others, creating new opportunities for social interaction and connection.

Enhanced Social Experiences: The metaverse can provide users with a more engaging and immersive social experience than traditional social networking platforms. Users can create avatars that represent themselves in the virtual world. They can explore virtual environments and engage in social activities and events with other users [58]. This can help users build stronger connections with others and create more meaningful social experiences. For example, users can attend virtual conferences, networking events, and trade shows where they can meet and interact with other users from around the world.

Increased Personalization: The metaverse can provide users with increased personalization options for their social experiences. Users can customize their avatars, virtual spaces, and social interactions to reflect their personalities and interests [59]. This can help users express themselves more fully and build stronger connections with others who share their interests. Users can design and decorate their virtual spaces, create and sell virtual goods, and engage in virtual commerce.

Global Accessibility: The metaverse can provide social networking opportunities to people who might not have access to traditional social networking platforms. The metaverse can be accessed from anywhere in the world, providing opportunities for users to connect with others from diverse backgrounds and cultures [60]. This can help users build global connections and expand their social networks. For example, users can attend virtual language exchanges, cultural events, and study groups.

New Social Activities: The metaverse can enable new and innovative social activities that are not possible in traditional social networking platforms [33]. Users can attend virtual concerts, parties, and events or play games with other users in a virtual environment. These activities can provide users a more engaging and immersive social experience and create new opportunities for social interaction and connection. For example, users can join virtual sports teams, book clubs, or gaming groups.

Shared Social Experiences: The metaverse can provide users with shared social experiences, allowing them to connect and engage with others in real-time, e.g., during traveling [61]. Users can explore virtual worlds and engage in social activities together, creating shared memories and experiences. This can help users build stronger connections with others and create a sense of community. For example, users can participate in virtual team-building exercises, group projects, and social impact initiatives.

Enhanced Privacy: The metaverse can provide users with enhanced privacy options for their social interactions [17]. Users can control who they interact with and choose to remain anonymous if they wish. This can provide users with a greater sense of security and control over their social interactions. Users can also participate in confidential support groups, sensitive discussions, or private events.

New Revenue Streams: Social networking companies can generate new revenue streams by leveraging the metaverse [62]. For instance, they can sell virtual goods and experiences, such as virtual clothing or concert tickets, to users [9, 54]. This can provide social networking companies with new opportunities for revenue generation and growth. Social networking companies can offer virtual sponsorship and advertisement opportunities, product placements, or affiliate marketing programs.

7.3.5 Healthcare

Metaverse has the potential to revolutionize healthcare by providing a platform for virtual healthcare services, medical education and training, healthcare research, mental health services, rehabilitation services, and health and wellness services [10]. As the metaverse continues to develop, we can expect to see new and innovative ways for healthcare providers to leverage the metaverse to improve patient care and outcomes.

Virtual Healthcare Services: The metaverse can be used to provide virtual healthcare services such as telemedicine, which allows patients to receive medical care remotely via video conferencing [63–65]. This can be particularly beneficial for patients who live in remote areas, have mobility or transportation issues, or need specialized care that is not available in their area. Additionally, virtual healthcare services can be more cost-effective and convenient for patients, as they can avoid travel and wait times.

The metaverse can also be used to provide other types of virtual healthcare services, such as health coaching, mental health counseling, and chronic disease management. For example, patients with diabetes could use the metaverse to receive virtual coaching on managing their condition, including personalized nutrition and exercise plans.

Medical Education: The metaverse can be used to provide medical education and training in a safe and controlled environment [66]. Medical students and healthcare professionals can use the metaverse to simulate medical scenarios and practice medical procedures before working with real patients. This can help them gain confidence and experience and reduce the risk of errors.

The metaverse can also be used to provide continuing education for healthcare professionals. For example, doctors and nurses could use the metaverse to attend virtual conferences and seminars as well as learn about the latest medical technologies and treatments.

Healthcare Research: The metaverse can facilitate healthcare research, including clinical trials and studies [63]. Researchers can use the metaverse to recruit study participants, collect data, and analyze results. This can help researchers conduct studies more efficiently and cost-effectively, as they can reach a larger and more diverse participant pool. Additionally, the metaverse can be used to simulate medical conditions and test new medical treatments.

Mental Health Services: The metaverse can be used to provide mental health services such as virtual therapy sessions, support groups, and mindfulness practices [63]. Mental health professionals can use the metaverse to provide virtual care services to patients who may be unable to attend in-person therapy sessions or who may prefer the privacy and convenience of virtual care. Additionally, the metaverse can be used to simulate anxiety-inducing scenarios to help patients practice coping strategies.

Rehabilitation: The metaverse can be used to provide rehabilitation services such as virtual physical therapy and cognitive training [67]. Patients can use the metaverse to simulate activities of daily living, practice exercises, and improve their cognitive skills. This can help patients recover from injuries or disabilities more quickly and effectively, as they can practice rehabilitation exercises in a safe and controlled environment.

Health and Wellness: The metaverse can be used to provide health and wellness services such as virtual fitness classes, nutrition coaching, and stress management [68]. Users can use the metaverse to engage in virtual workouts, track their fitness progress, and connect with others who share their health and wellness goals. This can help users improve their overall health and well-being, as they can access health and wellness services from anywhere, anytime.

7.3.6 Industrial Use

The industrial use of metaverse technology has the potential to revolutionize how businesses operate in various industries. By leveraging the power of virtual world technology, companies can improve efficiency, safety, collaboration, and customer engagement, leading to better outcomes and a more competitive edge in the marketplace. Here are some examples of Metaverse industrial uses:

Virtual Prototyping: Virtual prototyping is particularly useful for industries that require extensive product development, such as manufacturing, construction, and engineering [69, 70]. With metaverse technology, companies can create virtual models of their products, test different scenarios, and make adjustments before committing to physical production. This can save time and money, reduce waste, and improve the quality of the final product.

Virtual Training: Metaverse technology can be used for employee training in a variety of industries, including healthcare, aviation, and military [71, 72]. By simulating real-world scenarios in a virtual environment, employees can gain practical experience without the risk of injury or damage to equipment. This can improve safety, reduce costs associated with accidents and errors, and increase employee confidence and performance.

Remote Collaboration: Metaverse technology can facilitate remote collaboration between teams, regardless of their physical locations. This can be particularly useful for companies with distributed teams or global offices, as it can reduce travel costs and facilitate more efficient communication and collaboration between team members [73]. Metaverse technology can also provide a more engaging and immersive experience for remote workers, improving their sense of connection to the team and the company as a whole.

Marketing and Advertising: Metaverse technology can be used to create interactive and immersive marketing and advertising experiences for customers [52]. Companies can create virtual showrooms, product demos, and other experiences to showcase their products and services, providing a unique and engaging experience for potential customers. This can increase engagement and interest in products and services and provide valuable data on customer behavior and preferences.

Data Visualization: Metaverse technology can be used to visualize complex data sets in a virtual environment, making it easier for employees to understand and analyze data [71]. This can be particularly useful in industries such as finance, healthcare, and scientific research, where large amounts of data must be analyzed and interpreted. By visualizing data in a virtual environment, companies can gain insights that may be difficult to see in traditional data visualization tools.

Omniverse: The Omniverse is designed as a platform for creating and simulating virtual environments with high realism and interactivity. It is based on NVIDIA's technology, which is a graphics processing unit (GPU) that can handle advanced real-time rendering and simulation tasks [74].

The Omniverse is designed to be an open platform that allows users to create and share content, collaborate on projects, and connect with other creators and developers. It is also scalable and flexible, allowing it to be used for a wide range of applications, from architecture and engineering to film and entertainment.

One of the key features of the Omniverse is its support for real-time rendering and simulation [75]. This means that changes made to a virtual environment can be instantly visualized and tested, allowing creators to iterate and refine their designs quickly. The Omniverse also supports physics simulation, allowing objects to interact with each other and with the environment in a realistic and accurate way.

Another important feature of the Omniverse is its support for interoperability [76]. This means that content created in other software platforms can be imported into the Omniverse and vice versa, allowing creators to use the tools they are most comfortable with and to collaborate with others who may be using different software.

The Omniverse represents an exciting new direction in the development of virtual worlds and environments. Its emphasis on photorealism, physical accuracy, and real-time simulation has the potential to revolutionize a wide range of industries, from architecture and engineering to film and entertainment.

NVIDIA has also played a key role in advancing the field of generative AI. Generative AI involves using deep learning algorithms to generate new and original content similar to existing data. This is typically achieved by training a neural network on a large dataset of images, videos, music, or text and then using that network to generate new content similar in style or theme.

7.3.7 Entertainment

Metaverse has the potential to revolutionize the entertainment industry by providing new and innovative ways for creators to engage with audiences and monetize their work. By taking advantage of the unique features and capabilities of the metaverse, entertainment companies can create immersive and personalized user experiences, expand their distribution channels, and gain valuable data insights into user behavior and engagement.

Interactive and Immersive Experiences: The metaverse can provide users with a level of interactivity and immersion that is not possible in the physical world [77, 78]. VR technology can create a sense of presence and immersion that allows users to feel like they are actually in a different world. This can be particularly powerful for entertainment experiences such as concerts, where users can feel like they are actually in the audience, even if they are watching from home.

Social and Community Building: The metaverse can also provide opportunities for social and community building around entertainment experiences [78]. Users can connect with each other and form communities around shared interests, such as gaming, music, or film. For example, in the metaverse, users can attend virtual music festivals with their friends or interact with other fans of a particular band. This can create a sense of belonging and engagement for users and provide opportunities for creators to engage with their audiences in new and exciting ways.

New Revenue Streams: The metaverse can provide new revenue streams for entertainment companies and creators [62]. For example, virtual goods and merchandise, such as virtual clothing, accessories, and even virtual real estate [79], can be sold within the metaverse [79]. Additionally, creators can monetize their content by advertising or selling digital products such as music, films, and games. This can provide new and innovative ways for creators to earn revenue from their work and also opportunities for new and emerging artists to reach audiences.

New Distribution Channels: The metaverse can also provide new distribution channels for entertainment content. For example, films, music, and games can be distributed within the metaverse, providing users with new and innovative ways to consume content [80, 81]. Additionally, the metaverse can provide opportunities for niche content that may not be popular enough for traditional distribution channels but can find an audience within the metaverse. For example, VR gaming experiences that are too niche for mainstream audiences could find a following in the metaverse.

Gaming: Gaming is one of the most obvious ways the metaverse can benefit entertainment [79]. The metaverse can provide users with new and innovative gaming experiences beyond what is possible in the physical world. Games can be developed specifically for the metaverse, taking advantage of the unique features and capabilities of the platform. Additionally, gaming communities can form within the metaverse, providing opportunities for socializing and competition. For example, users can compete in VR sports or play games with their friends worldwide.

Fan Engagement: The metaverse can provide new ways for fans to engage with their favorite entertainment properties. For example, fans can attend virtual events such as movie premieres, book signings, or Q&A sessions with creators

[82]. They can also interact with virtual versions of their favorite characters or explore virtual worlds inspired by their favorite movies, books, or TV shows. This can create a deeper level of engagement and connection between fans and their favorite properties.

Personalization: The metaverse can also provide personalized entertainment experiences for users [43]. A virtual assistant or AI could learn about user's preferences and recommend content or experiences they would enjoy. This could help users discover new entertainment properties that they may not have otherwise found. It could also provide a more tailored and satisfying experience.

Accessibility: The metaverse can also provide opportunities for more accessible entertainment experiences. VR technology can provide experiences that are accessible to people with physical disabilities who may not be able to attend live events or participate in certain activities in the physical world [83]. Additionally, the metaverse can provide language translation or closed captioning for users with difficulty understanding content in its original language.

New Creative Opportunities: The metaverse can also provide new creative opportunities for entertainment creators [84]. VR technology can allow creators to experiment with new forms of storytelling and visual effects that may not be possible in the physical world. Additionally, the metaverse can provide opportunities for cross-platform integration, allowing creators to connect multiple forms of entertainment, such as music, film, and gaming, into a seamless and immersive experience.

Data Insights: Finally, the metaverse can provide valuable data insights for companies and creators. User behavior and engagement data can help companies understand what content and experiences are most popular with audiences or customers and can inform future creative decisions [27, 85]. Additionally, user data can help companies understand the most effective advertising and marketing strategies in the metaverse.

7.4 Accelerating Killer Apps

The semiconductor industry can support metaverse applications by providing the hardware, components, and technology required to power and enable immersive, interactive, and secure metaverse experiences. As metaverse applications continue to evolve and grow, the role of the semiconductor industry in supporting these applications is likely to become even more critical. The following are ways that the semiconductor industry can support metaverse applications:

Powering the Hardware: The metaverse relies heavily on devices such as smartphones, tablets, PCs, and gaming consoles, which all require significant processing power to run [86]. The semiconductor industry can provide the necessary microprocessors, GPUs, and other components required to power these devices. Additionally, the semiconductor industry can also develop energy-efficient chips that consume less power and generate less heat, which is particularly important for mobile devices that run on battery power.

Advancing Chip Technology: As metaverse applications become more complex and sophisticated, the demand for powerful and efficient chips will continue to grow. The semiconductor industry can continue to advance chip technology by developing smaller, faster, and more energy-efficient chips [86]. The industry can develop new manufacturing techniques that allow the creation of chips with more transistors, resulting in faster processing speeds and better performance.

Providing Sensors and Components: Metaverse applications often rely on a variety of sensors and components to deliver immersive and interactive experiences [87]. The semiconductor industry can provide these components, such as cameras, microphones, haptic feedback sensors, and motion-tracking sensors, to enable a wide range of metaverse applications.

Developing Specialized Chips: The semiconductor industry can also develop specialized chips that are optimized for specific metaverse applications [88]. The industry can develop chips that are specifically designed for machine learning or computer vision, which are critical components of many metaverse applications. These specialized chips can enhance the performance and capabilities of metaverse applications, resulting in a more immersive and engaging user experience.

Ensuring Security: Metaverse applications are likely to handle sensitive data, such as personal and financial data. Therefore, it is crucial to ensure the security of these applications. The semiconductor industry can develop security chips to protect sensitive data and prevent unauthorized access. These chips can incorporate encryption and other security features that help keep user data safe. For example, Ref. [89] uses a security chip for an authentication protocol.

7.4.1 Edge Computing

Edge computing can provide several benefits to the metaverse, including reduced latency, improved scalability, enhanced security, personalized experiences, and cost-efficiency. As the metaverse continues to evolve, we can expect to see more integration between edge computing and metaverse technologies, leading to even more immersive and engaging user experiences [90, 91]. Here, we look at key areas where edge computing can benefit the metaverse.

Reduced Latency: In addition to processing graphics locally on the user's device, edge computing can also be used to process other types of data closer to the source [16, 92, 93]. Edge devices can be used to process user inputs, such as movement and actions in the virtual environment, and transmit the data to the centralized server for processing. This can reduce the lag time between the user's actions and the response from the system, leading to a more immersive and responsive experience.

Scalability: Edge computing can also help the metaverse scale by distributing storage and processing across multiple edge devices [26]. Edge devices can store and process user-generated content, such as virtual objects and environments, reducing the load on centralized servers. This can help the metaverse support more users and applications without sacrificing performance or reliability.

Security: In addition to processing sensitive data locally on the user's device, edge computing can also be used to encrypt and secure data as it is transmitted over the network. This can help prevent data breaches and cyber-attacks, as well as protect user privacy [26]. Edge devices can be used to encrypt data before it is transmitted to the centralized server, ensuring that sensitive data is always protected.

Personalization: Edge computing can also enable more personalized experiences in the metaverse by processing user data locally on the device [26]. Edge devices can be used to process user preferences, such as music or entertainment preferences, and provide personalized recommendations and content. This can enhance the user experience and increase user engagement.

Cost-Efficiency: In addition to reducing the costs associated with centralized computing, edge computing can also provide cost savings on bandwidth and network usage [94]. Edge devices can compress and optimize data before it is transmitted over the network, reducing the amount of data that needs to be transmitted and the network usage. This can lead to cost savings on bandwidth and network usage. It can also improve the overall system performance.

7.4.2 Acceleration by Specialized Hardware

The combination of GPU, tensor processing unit (TPU), data processing unit (DPU), field programmable gate array (FPGA), and systolic array chips can help to power the metaverse and enable more advanced and immersive user experiences. By providing specialized processing capabilities, these chips can help to handle the massive amounts of data and computation required by metaverse applications while also improving the overall performance and interactivity of the experience.

GPUs can accelerate the processing of graphics and visual effects required by metaverse applications [73, 95, 96]. They are designed to handle massive amounts of parallel processing, making them ideal for applications that require complex visual rendering, such as virtual worlds, AR applications, and 3D modeling.

TPUs can accelerate the processing of AI and machine learning applications required by metaverse applications [95, 96]. They are designed to handle large-scale matrix operations, making them ideal for applications that require natural language processing, speech recognition, and image and video recognition.

DPUs can accelerate the processing of large datasets required by metaverse applications. DPUs are designed to handle complex data processing tasks, making them ideal for applications that require real-time sensor data processing, user behavior analysis, and physics simulations [96].

FPGAs are programmable chips that can be configured to perform a wide range of tasks. In the metaverse, FPGA chips can be used to implement specialized algorithms or process pipelines that are unique to a particular application. FPGAs can be used to accelerate specific parts of a metaverse application, such as physics simulations or rendering certain types of graphics [97].

Systolic arrays are a type of specialized hardware architecture for matrix multiplication and other matrix-based operations. They are highly parallel and can quickly process large amounts of data, making them ideal for applications that require complex matrix operations, such as AI and machine learning [98].

References

1 Stephenson, N. (2003). *Snow Crash: A Novel.* Spectra.

2 Al-Ghaili, A.M., Kasim, H., Al-Hada, N.M. et al. (2022). A review of metaverse's definitions, architecture, applications, challenges, issues, solutions, and future trends. *IEEE Access* 10: 125835–125866.

3 Lee, L.H., Braud, T., Zhou, P. et al. (2021). All one needs to know about metaverse: a complete survey on technological singularity, virtual ecosystem, and research agenda. *arXiv preprint arXiv: 2110.05352.*

4 Mystakidis, S. (2022). Metaverse. *Encyclopedia* 2 (1): 486–497.

5 Sami, H., Hammoud, A., Arafeh, M. et al. (2023). The metaverse: survey, trends, novel pipeline ecosystem and future directions. *arXiv preprint arXiv:2304.09240.*

6 Mozumder, M.A.I., Sheeraz, M.M., Athar, A. et al. (2022). Overview: technology roadmap of the future trend of metaverse based on IoT, blockchain, AI technique, and medical domain metaverse activity. *2022 24th International Conference on Advanced Communication Technology (ICACT)*, 13–16 February 2022, Pyeongchang, South Korea, pp. 256–261. IEEE.

7 Singh, H., Gupta, S., and Hassan, M. (2022). Metaverse: a new digital marketing trend. *International Journal of Multidisciplinary Research and Analysis* 5 (12): 3623–3628.

8 Jaynes, C., Seales, W.B., Calvert, K. et al. (2003). The metaverse: a networked collection of inexpensive, self-configuring, immersive environments.

Proceedings of the Workshop on Virtual Environments 2003, 22–23 May 2003, Zurich Switzerland, pp. 115–124.

9 Rathore, B. (2023). Digital transformation 4.0: integration of artificial intelligence & metaverse in marketing. *Eduzone: International Peer Reviewed/Refereed Multidisciplinary Journal* 12 (1): 42–48.

10 Lee, C.W. (2022). Application of metaverse service to healthcare industry: a strategic perspective. *International Journal of Environmental Research and Public Health* 19 (20): 13038.

11 Mistretta, S. (2022). The metaverse – an alternative education space. In: *AI, Computer Science and Robotics Technology*. IntechOpen https://doi.org/10.5772/acrt.05.

12 Sharma, R.C., Sharma, Y.P., Designing virtual reality experiences in education. *Bulletin of the Technical Committee on Learning Technology*, 21(1): 19–22 March 2021.

13 Wiederhold, B.K. and Riva, G. (2022). Metaverse creates new opportunities in healthcare. *Annual Review of CyberTherapy and Telemedicine* 20: 3–7.

14 Zhang, X., Chen, Y., Hu, L., and Wang, Y. (2022). The metaverse in education: definition, framework, features, potential applications, challenges, and future research topics. *Frontiers in Psychology* 13: https://doi.org/10.3389/fpsyg.2022.1016300.

15 Naqvi, N. (2023). Metaverse for public good: embracing the societal impact of metaverse economies. *The Journal of the British Blockchain Association* https://doi.org/10.31585/jbba-6-1-(6)2023 (accessed on 25 May 2023).

16 Maksymyuk, T., Gazda, J., Bugár, G. et al. (2022). Blockchain-empowered service management for the decentralized metaverse of things. *IEEE Access* 10: 99025–99037.

17 Maloney, D., Zamanifard, S., and Freeman, G. (2020). Anonymity vs. familiarity: self-disclosure and privacy in social virtual reality. *Proceedings of the 26th ACM Symposium on Virtual Reality Software and Technology*, 1–4 November 2020, Virtual Conference, pp. 1–9.

18 Cheng, S., Zhang, Y., Li, X. et al. (2022). Roadmap toward the metaverse: an AI perspective. *The Innovation* 3 (5): 100293.

19 Guo, Y., Yu, T., Wu, J. et al. (2022). Artificial intelligence for metaverse: a framework. *CAAI Artificial Intelligence Research* 1 (1): 54–67.

20 Thien, H.Y., Pham, Q.V., Pham, X.Q. et al. (2023). Artificial intelligence for the metaverse: a survey. *Engineering Applications of Artificial Intelligence* 117: 105581.

21 Cao, L. (2022). Decentralized AI: edge intelligence and smart blockchain, metaverse, web3, and desci. *IEEE Intelligent Systems* 37 (3): 6–19.

22 Liu, H., Wang, Y., Fan, W. et al. (2022). Trustworthy AI: a computational perspective. *ACM Transactions on Intelligent Systems and Technology* 14 (1): 1–59.

23 Golatowski, F. et al. (2019). Challenges and research directions for blockchains in the Internet of Things. *2019 IEEE International Conference on Industrial Cyber Physical Systems (ICPS),* 6–9 May 2019, Taipei, pp. 712–717.

24 Hwang, G.J. and Chien, S.Y. (2022). Definition, roles, and potential research issues of the metaverse in education: an artificial intelligence perspective. *Computers and Education: Artificial Intelligence* 3: 100082.

25 Yang, Q., Zhao, Y., Huang, H. et al. (2022). Fusing blockchain and AI with metaverse: a survey. *IEEE Open Journal of the Computer Society* 3: 122–136.

26 Xu, M., Ng, W.C., Lim, W.Y.B. et al. (2022). A full dive into realizing the edge-enabled metaverse: visions, enabling technologies, and challenges. *IEEE Communications Surveys & Tutorials* 25 (1): 656–700.

27 Dwivedi, Y.K., Hughes, L., Wang, Y. et al. (2023). Metaverse marketing: how the metaverse will shape the future of consumer research and practice. *Psychology & Marketing* 40 (4): 750–776.

28 Hilken, T., Chylinski, M., de Ruyter, K. et al. (2022). Exploring the frontiers in reality-enhanced service communication: from augmented and virtual reality to neuro-enhanced reality. *Journal of Service Management* 33 (4/5): 657–674.

29 Gadekallu, T.R., Huynh-The, T., Wang, W. et al. (2022). Blockchain for the metaverse: a review. *arXiv preprint arXiv:2203.09738.*

30 Xu, H., Li, Z., Li, Z. et al. (2022). Metaverse native communication: a blockchain and spectrum prospective. *2022 IEEE International Conference on Communications Workshops (ICC Workshops),* 16–20 May 2022, Seoul South Korea, pp. 7–12.

31 Dorri, A., Luo, F., Karumba, S. et al. (2021). Temporary immutability: a removable blockchain solution for prosumer-side energy trading. *Journal of Network and Computer Applications* 180: 103018.

32 Cai, Y., Llorca, J., Tulino, A.M., and Molisch, A.F. (2022). Compute-and data-intensive networks: the key to the metaverse. *2022 1st International Conference on 6G Networking (6GNet),* 6–8 July 2022, Paris France, pp. 1–8. IEEE.

33 George, A.H., Fernando, M., George, A.S. et al. (2021). Metaverse: the next stage of human culture and the internet. *International Journal of Advanced Research Trends in Engineering and Technology (IJARTET)* 8 (12): 1–10.

34 Ning, H., Wang, H., Lin, Y. et al. (2021). A survey on metaverse: the state-of-the-art, technologies, applications, and challenges. *arXiv preprint arXiv:2111.09673.*

35 Spajić, J., Mitrović, K., Lalić, D. et al. (2022). Personalized brand experience in metaverse. *10th International Conference on Mass Customization and*

Personalization–Community of Europe (MCP-CE 2022), 21–23 September 2022, Novi Sad Serbia, pp. 21–23.

36 Han, Y., Niyato, D., Leung, C. et al. (2022). A dynamic resource allocation framework for synchronizing metaverse with IoT service and data. *ICC 2022-IEEE International Conference on Communications*, 16–20 May 2022, Seoul South Korea, pp. 1196–1201.

37 Torres Vega, M., Liaskos, C., Abadal, S. et al. (2020). Immersive interconnected virtual and augmented reality: a 5G and IoT perspective. *Journal of Network and Systems Management* 28: 796–826.

38 Dai, N., Lei, I.M., Li, Z. et al. (2023). Recent advances in wearable electromechanical sensors – moving towards machine learning-assisted wearable sensing systems. *Nano Energy* 105: 108041.

39 Guan, J., Irizawa, J., and Morris, A. (2022). Extended reality and internet of things for hyper-connected metaverse environments. *2022 IEEE Conference on Virtual Reality and 3D User Interfaces Abstracts and Workshops (VRW)*, 12–16 March 2022, Virtual Conference, pp. 163–168. IEEE.

40 Jo, D. and Kim, G.J. (2019). IoT+ AR: pervasive and augmented environments for "Digi-log" shopping experience. *Human-Centric Computing and Information Sciences* 9 (1): 1–17.

41 Duan, H., Li, J., Fan, S. et al. (2021). Metaverse for social good: a university campus prototype. *Proceedings of the 29th ACM International Conference on Multimedia*, 20–24 October 2021, Virtual Conference, pp. 153–161.

42 Allam, Z., Sharifi, A., Bibri, S.E. et al. (2022). The metaverse as a virtual form of smart cities: opportunities and challenges for environmental, economic, and social sustainability in urban futures. *Smart Cities* 5 (3): 771–801.

43 Yemenici, A.D. (2022). Entrepreneurship in the world of metaverse: virtual or real? *Journal of Metaverse* 2 (2): 71–82.

44 Jungherr, A. and Schlarb, D.B. (2022). The extended reach of game engine companies: how companies like epic games and unity technologies provide platforms for extended reality applications and the metaverse. *Social Media+Society* 8 (2): 1–12.

45 Contreras, G.S., González, A.H., Fernández, M.I.S. et al. (2022). The importance of the application of the metaverse in education. *Modern Applied Science* 16 (3): 1–34.

46 Rahman, K.R., Shitol, S.K., Islam, M.S. et al. (2023). Use of metaverse technology in education domain. *Journal of Metaverse* 3 (1): 79–86.

47 Dahan, N.A., Al-Razgan, M., Al-Laith, A. et al. (2022). Metaverse framework: a case study on E-learning environment (ELEM). *Electronics* 11 (10): 1616.

48 Joshi, S. and Pramod, P.J. (2023). A collaborative metaverse based A-La-carte framework for tertiary education (CO-MATE). *Heliyon* 9 (2): e13424.

49 De Freitas, S., Rebolledo-Mendez, G., Liarokapis, F. et al. (2010). Learning as immersive experiences: using the four-dimensional framework for designing and evaluating immersive learning experiences in a virtual world. *British Journal of Educational Technology* 41 (1): 69–85.

50 Barráez-Herrera, D.P. (2022). Metaverse in a virtual education context. *Metaverse* 3 (1): 1–9.

51 MacCallum, K. and Parsons, D. (2019). Teacher perspectives on mobile augmented reality: the potential of metaverse for learning. *World Conference on Mobile and Contextual Learning*, 16 September 2019, Delft Netherlands, pp. 21–28.

52 Nalbant, K.G. (2022). Development of digital marketing, advertising and sales strategies in the world of metaverse. *Academic Studies in Social, Human and Administrative Science* 161–178.

53 Visconti, R.M. (2022). From physical reality to the metaverse: a multilayer network valuation. *Journal of Metaverse* 2 (1): 16–22.

54 Jeong, H., Yi, Y., and Kim, D. (2022). An innovative e-commerce platform incorporating metaverse to live commerce. *International Journal of Innovative Computing, Information and Control* 18 (1): 221–229.

55 Mott, M., Tang, J., Kane, S. et al. (2020). "I just went into it assuming that I wouldn't be able to have the full experience" understanding the accessibility of virtual reality for people with limited mobility. *Proceedings of the 22nd International ACM SIGACCESS Conference on Computers and Accessibility*, 26–28 October 2020, Virtual Conference Greece, pp. 1–13.

56 Gursoy, D., Malodia, S., and Dhir, A. (2022). The metaverse in the hospitality and tourism industry: an overview of current trends and future research directions. *Journal of Hospitality Marketing & Management* 31 (5): 527–534.

57 Periyasami, S. and Periyasamy, A.P. (2022). Metaverse as future promising platform business model: case study on fashion value chain. *Businesses* 2 (4): 527–545.

58 Riva, G. and Wiederhold, B.K. (2022). What the metaverse is (really) and why we need to know about it. *Cyberpsychology, Behavior, and Social Networking* 25 (6): 355–359.

59 Freeman, G. and Maloney, D. (2021). Body, avatar, and me: the presentation and perception of self in social virtual reality. *Proceedings of the ACM on Human-Computer Interaction* 4 (CSCW3): 1–27.

60 Henz, P. (2022). The societal impact of the metaverse. *Discover Artificial Intelligence* 2 (1): 19.

61 Buhalis, D., Lin, M.S., and Leung, D. (2022). Metaverse as a driver for customer experience and value co-creation: implications for hospitality and tourism management and marketing. *International Journal of Contemporary Hospitality Management* 35 (2): 701–716.

62 Kraus, S., Kanbach, D.K., Krysta, P.M. et al. (2022). Facebook and the creation of the metaverse: radical business model innovation or incremental transformation? *International Journal of Entrepreneurial Behavior & Research* 28 (9): 52–77.

63 Chengoden, R., Victor, N., Huynh-The, T. et al. (2023). Metaverse for healthcare: a survey on potential applications, challenges and future directions. *IEEE Access* 11: 12765–12795.

64 Shu, S. and Woo, B.K.P. (2023). Pioneering the metaverse: the role of the metaverse in an aging population. *JMIR Aging* 6: e40582.

65 Wang, G., Badal, A., Jia, X. et al. (2022). Development of metaverse for intelligent healthcare. *Nature Machine Intelligence* 4 (11): 1–8.

66 Petrigna, L. and Musumeci, G. (2022). The metaverse: a new challenge for the healthcare system: a scoping review. *Journal of Functional Morphology and Kinesiology* 7 (3): 63.

67 Yeung, A.W.K., Tosevska, A., Klager, E. et al. Augmented reality applications in medicine: analysis of the scientific literature. *Journal of Medical Internet Research* 23 (2): e25499.

68 Thomason, J. (2021). Metahealth-how will the metaverse change health care? *Journal of Metaverse* 1 (1): 13–16.

69 Berg, L.P. and Vance, J.M. (2017). Industry use of virtual reality in product design and manufacturing: a survey. *Virtual Reality* 21: 1–17.

70 Shi, Q., Dong, B., He, T. et al. (2020). Progress in wearable electronics/photonics – moving toward the era of artificial intelligence and internet of things. *InfoMat* 2 (6): 1131–1162.

71 Siyaev, A. and Jo, G.S. (2021). Towards aircraft maintenance metaverse using speech interactions with virtual objects in mixed reality. *Sensors* 21 (6): 2066.

72 Mitra, S. (2023). Metaverse: a potential virtual-physical ecosystem for innovative blended education and training. *Journal of Metaverse* 3 (1): 66–72.

73 Balzerkiewitz, H.P. and Stechert, C. (2020). Use of virtual reality in product development by distributed teams. *Procedia CIRP* 91: 577–582.

74 NVIDIA Omniverse Home Page: The platform for developing OpenUSD applications for industrial digitalization and generative physical AI. NVIDIA 2024. https://www.nvidia.com/en-us/omniverse/ (accessed on 1 September 2024).

75 Hummel, M. and van Kooten, K. (2019). Leveraging NVIDIA omniverse for in situ visualization. *High Performance Computing: ISC High Performance 2019 International Workshops,* 16–20 June 2019, *Frankfurt Germany, Revised Selected Papers 34,* pp. 634–642. Springer International Publishing.

76 Chamusca, I.L., Santos, F.V.D.J., Ferreira, C.V. et al. (2022). Evaluation of design guidelines for the development of intuitive virtual reality authoring tools: a case study with NVIDIA omniverse. *2022 IEEE International Symposium on Mixed and Augmented Reality Adjunct (ISMAR-Adjunct)*, 17–21 October 2022, Singapore, pp. 357–362. IEEE.

77 Kirkpatrick, K. (2022). Applying the metaverse. *Communications of the ACM* 65 (11): 16–18.

78 Han, D.I.D., Bergs, Y., and Moorhouse, N. (2022). Virtual reality consumer experience escapes: preparing for the metaverse. *Virtual Reality* 1–16.

79 Hollensen, S., Kotler, P., and Opresnik, M.O. (2023). Metaverse – the new marketing universe. *Journal of Business Strategy* 44 (3): 119–125.

80 Jia, T., Wang, Y., and Cho, N.G. (2022). A new model of dance performance art – performance analysis of metaverse application. *Art and Society* 1 (3): 44–51.

81 Liu, J. (2022). Metaverse and brand: a study of luxury brand digital marketing strategy-taking gucci as an example. *2022 4th International Conference on Economic Management and Cultural Industry (ICEMCI 2022)*, 14–16 October 2022, Virtual Conference Chongqing China, pp. 1907–1913. Atlantis Press.

82 Kye, B., Han, N., Kim, E. et al. (2021). Educational applications of metaverse: possibilities and limitations. *Journal of Educational Evaluation for Health Professions* 18: 32.

83 Seigneur, J.M. and Choukou, M.A. (2022). How should metaverse augment humans with disabilities? *13th Augmented Human International Conference*, 26–27 May 2022, Winnipeg MB Canada, pp. 1–6.

84 Hazan, E., Kelly, G., Khan, H., Spillecke, D., and Yee, L. (2022). Marketing in the metaverse: an opportunity for innovation and experimentation. *The McKinsey Quarterly*. https://www.mckinsey.com/capabilities/growth-marketing-and-sales/our-insights/marketing-in-the-metaverse-an-opportunity-for-innovation-and-experimentation#/ (accessed on 20 May 2023).

85 Adams, D. (2022). Virtual retail in the metaverse: customer behavior analytics, extended reality technologies, and immersive visualization systems. *Linguistic and Philosophical Investigations* 21: 73–88.

86 Gottscho, R.A., Levine, E.V., Liu, T.J.K. et al. (2022). Innovating at speed and at scale: a next generation infrastructure for accelerating semiconductor technologies. *arXiv preprint arXiv:2204.02216*.

87 Park, S.M. and Kim, Y.G. (2022). A metaverse: taxonomy, components, applications, and open challenges. *IEEE Access* 10: 4209–4251.

88 Braud, T., Lee, L.H., Alhilal, A. et al. (2022). Dios-an extended reality operating system for the metaverse. *arXiv preprint arXiv:2201.03256*.

89 Zhang, X., Huang, X., Yin, H. et al. (2022). Llakep: a low-latency authentication and key exchange protocol for energy internet of things in the metaverse era. *Mathematics* 10 (14): 2545.

90 Dhelim, S., Kechadi, T., Chen, L. et al. (2022). Edge-enabled metaverse: the convergence of metaverse and mobile edge computing. *arXiv preprint arXiv:2205.02764.*

91 Yu, J., Alhilal, A., Hui, P., and Tsang, D.H. (2022). 6G mobile-edge empowered metaverse: requirements, technologies, challenges and research directions. *arXiv preprint arXiv:2211.04854.*

92 Wang, Y. and Zhao, J. (2022). A survey of mobile edge computing for the metaverse: architectures, applications, and challenges. *arXiv preprint arXiv:2212.00481.*

93 Karunarathna, S., Wijethilaka, S., Ranaweera, P. et al. (2023). The role of network slicing and edge computing in the metaverse realization. *IEEE Access* 11: 25502–25530.

94 Cao, K., Liu, Y., Meng, G., and Sun, Q. (2020). An overview on edge computing research. *IEEE Access* 8: 85714–85728.

95 Zhang, J. and Tao, D. (2020). Empowering things with intelligence: a survey of the progress, challenges, and opportunities in artificial intelligence of things. *IEEE Internet of Things Journal* 8 (10): 7789–7817.

96 Yao, J., Zhang, S., Yao, Y. et al. (2022). Edge-cloud polarization and collaboration: a comprehensive survey for AI. *arXiv preprint arXiv:2111.06061.*

97 Nguyen, T. (2020). Low-latency mixed reality headset. *Low-Latency VR/AR Headset Project from Conix Research Center, Computing on Network Infrastructure for Pervasive Perception, Cognition and Action.* https://people.eecs.berkeley.edu/~kubitron/courses/cs262a-F19/projects/reports/project14_report.pdf (accessed on 24 May 2023).

98 Kung, H.T., McDanel, B., and Zhang, S.Q. (2019). Packing sparse convolutional neural networks for efficient systolic array implementations: column combining under joint optimization. *Proceedings of the Twenty-Fourth International Conference on Architectural Support for Programming Languages and Operating Systems*, 13–17 April 2019, Providence RI USA, pp. 821–834.

8

From Concept to Production

8.1 System Design Process

In this chapter, we will consider some of the fundamental engineering aspects of making a product from concept to production. There is not much point in developing a new product unless it fully delivers on its value proposition through correct and reliable operation. There can be many pitfalls along the way, so taking the time upfront to apply the right validation and debug methodology can make all the difference in determining whether a product will ultimately enjoy success in the market.

To begin, let us presume that the team has defined a Marketing Requirements Document (MRD), Product Feature Specification, or a similar document that comprehensively describes the intended features of the product. A well-written specification should clearly state which industry specifications the product complies with, explain the various functions and capabilities, define the operating conditions under which the device can be used, and state performance and power consumption targets. The items above can be considered direct functionalities; the indirect aspects related to provisioning, deployment, diagnostics, operation, maintenance, security, quality, and reliability are equally important. These topics are often given less emphasis up front and run the risk of becoming the oversight that gates production or impacts sales. For example, a smart network interface card (SmartNIC) device that delivers excellent performance and supports impactless firmware updates but lacks an anti-rollback feature may fail to win market adoption due to the security risk it presents to the datacenter.

Supposing we are suitably armed with a robust MRD and associated technical specifications, and these requirements have been carefully broken down into sub-features across domains, the engineering team should be able to proceed with the design and implementation of the hardware, firmware, and software components. During this phase, it is crucial for the validation process [1, 2] to

Edge Computing Acceleration: From 5G to 6G and Beyond, First Edition.
Patrick Hung, Hongwei Kan, and Greg Knopf.

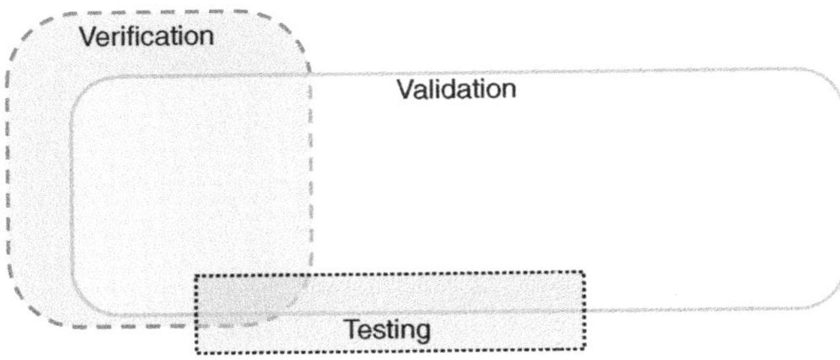

Figure 8.1 Key activities ensuring final products meet MRD and other specifications: verification, validation, and testing.

begin in earnest, even though there may be many months or quarters before the first prototypes arrive in the validation labs (Figure 8.1).

Before delving further into this topic, let us differentiate between verification, validation, and testing. These disciplines have very different and distinct purposes, though to some extent overlapping activities.

- **Verification:** Verification, also sometimes referred to as simulation or pre-silicon validation, refers to the use of electronic design automation (EDA) software to verify the design correctness through simulation of design behavior under various stimuli and checking the correctness of the behavior [3, 4]. This is typically done through a combination of unit-level tests for various functional unit blocks (also known as "IPs") and higher-level tests involving clusters of IPs or all IPs (full chip). Depending on the product in question, the verification plan may include gate-level simulation (GLS), analog mixed-signal verification, formal equivalence verification, and emulation (loading the model into Field-Programmable Gate Arrays (FPGAs) for faster execution of verification tests).
- **Validation:** This is a post-silicon discipline of ensuring that the product meets the specification [5, 6]. This encompasses the development of test plans along with creating the requisite hardware, software, and tools to execute the plans. We explore this topic in further detail below.
- **Testing:** Or more specifically, "manufacturing test" – refers to the activity of testing each specific unit during the process of manufacturing to ensure that it is functional prior to shipping it to a customer [7, 8]. The equipment, methodology, and governing factors in manufacturing tests are very different from validation, particularly because of the constraints and economics. Each unit must be tested as quickly as possible to minimize the cost overhead, which ultimately takes away from the profit margin.

Validation planning must comprehend all of the disciplines above to create a thorough coverage plan that is also efficient to execute. The cost of finding a bug increases exponentially with each subsequent phase. For example, an RTL logic bug found during verification is relatively straightforward to fix, often requiring a brief debug effort, minor code changes, and a regression test to verify the fix. This sequence may often be completed by a single engineer in a day. The design and simulation environment provides the developer with complete visibility and control to determine the root cause and fix the bug. However, finding the same bug in post-silicon may be far more costly and time-consuming. The bug may not manifest in a straightforward fashion under normal operating conditions. There may be very little ability to control and observe the related logic for debugging. It may be difficult to exercise the affected logic in a predictable manner, and reproducing a failure reliably enough to debug it may require modifying firmware or software. Observing the errant behavior may require the use of specialized debug equipment such as JTAG probes, protocol analyzers, and oscilloscopes. After establishing a root cause hypothesis, it must be proven in the lab through some additional experiments, which typically involve running additional simulations to confirm that the behavior is aligned with the root cause hypothesis. As a result, debugging the issue in post-silicon is generally an order of magnitude more complex, typically involving many more people as the work spans multiple disciplines and therefore consuming far more time and resources [9].

After the root-cause of a problem has been confirmed, the topic of fixing or working around the bug must be addressed. In our first example, where the bug is found in simulation, fixing the problem is often as simple as changing a single line of RTL code in the design. However, in the post-silicon scenario, a design change is a large undertaking. A modern CPU (Central Processing Unit), APU (Accelerated Processing Unit), or ASIC (Application-Specific Integrated Circuit) is fabricated in multiple metal and silicon layers. In the simplest case, where a metal-only engineering change order (ECO) can be defined to fix the issue, the design change may be accomplished in a few months, though at a considerable cost associated with new masks and samples. If the change is in the silicon base layers, the cost is typically much higher, and the process is much lengthier. Base layer changes typically require resynthesis of the associated logic, timing verification, functional verification, new mask sets, manufacturing significant quantities of new samples, and post-silicon validation to confirm the positive impact of the change and confirm that no negative side effects were introduced.

Devising a firmware or software-based workaround is therefore highly preferred in most cases if it is possible to do so. For this reason, there has been a steady trend in the industry to make designs more configurable and controllable through a combination of many techniques.

Table 8.1 Comparison between two hardware design options.

	Design options	
	Hardcoded state machine logic	**MCU firmware control**
Design cost	Higher	Lower
Unit cost	Lower	Higher
Flexibility	Worse	Better
Risk	More	Less

Option 1: State machine logic design and Option 2: MCU firmware control design.

For example, consider complex flows such as training a high-speed electrical interface or orchestrating power management transitions across a system on a chip (SoC) [10–12]. These flows could be completely implemented in hardware through state machines and some control. Visibility may be provided through register settings and design for debug (DFD) facilities [13]. Alternatively, the logic may be partitioned into a set of fixed functions and a microcontroller which interfaces with the rest of the SoC and controls the functional logic. In this model, the protocol between the SoC host logic and the microcontroller can be used to initiate the various flows supported by the subsystem and carried out by the microcontroller firmware. While the additional complexity of this architecture may drive up more engineering costs, the flexibility it affords often far outweighs the cost (Table 8.1).

Supposing that we have such an architecture in place and we encounter the bug in post-silicon, there is a reasonable chance that a fix or workaround can be devised and implemented in firmware. This is typically the best-case scenario for resolving a bug discovered in post-silicon. However, this seemingly ideal approach comes with a hidden cost if overused in the form of compounding side effects [14]. Each workaround may have some individual side effects, such as changing timing, consuming resources, impacting performance, or altering behavior relative to specifications [15]. While a small number of workarounds applied simultaneously may not cause any issues, supporting dozens or hundreds in a given subsystem may ultimately be worse than simply fixing the design in the first place. Engineering teams should be mindful of these trade-offs throughout the development cycle.

Just as the debug process is more complex in post-silicon, we can see that the process of fixing or working around a bug in post-silicon during the development phase tends to be far more costly than fixing the same bug in pre-silicon [16].

Going one step further, let us suppose that the same bug is only found after the product has launched and deployed in the market. Reports of the issue coming in from the field will typically be based on very limited data [17] and may be accompanied by a high degree of urgency and anxiety as customers grapple with the impact the issue presents to their business operations. Sales, supplier and client relationships, executive relationships, performance to total cost of ownership (TCO) models or key performance indicators (KPIs), operating expenditures, and brand image can all suffer from the impact of a field issue. These factors amplify the importance of solving the problem quickly and completely with minimal burden to customers.

Access is often limited both physical and networked, and both the data gathering and conducting experiments are usually severely limited by policies and technical limitations. For example, a 5G edge server may be deployed in a remote location that is extremely difficult to access and subject to a range of corporate and governmental policies and regulations from the nation in which it resides. These corporate and regulatory limitations commonly prevent information sharing about how the server was being used at the time of the issue. They may also prohibit sharing log files or other details required to debug and fix the problem. Problems may also result in violating contractual terms and conditions, which can trigger costly remedies. For all these reasons and more, discovering the same hypothetical bug after the product has launched and been deployed widely in the field can easily cost millions or billions of dollars to remedy. In extreme scenarios, a single problem discovered in the field could become an existential threat to a business. The famous FDIV bug discovered by a professor in 1994 in the Intel Pentium processor is a noteworthy example, as it triggered a recall of CPUs and cost the company roughly $475M US dollars [18].

Now that we have considered the overwhelming benefits of finding and fixing bugs as early as possible in the product development lifecycle, let us consider how to approach validation planning to accomplish this objective. A robust validation plan should, at a minimum, consist of the following elements:

- **Breadth of Coverage:** This refers to the extent to which the validation plan covers various aspects of the product. It ensures that all features, functions, and scenarios are tested to catch potential issues. A comprehensive breadth of coverage helps in identifying edge cases and ensuring that the system behaves correctly under different conditions.
- **Depth of Coverage:** While breadth of coverage looks at the range of features and scenarios tested, depth of coverage refers to the level of detail and thoroughness in testing each scenario. It involves testing different permutations and combinations, exploring boundary conditions, and conducting targeted stress testing to ensure robustness.
- **Standards Compliance:** This element ensures that the system or product meets relevant industry standards, regulations, and guidelines. Compliance is crucial for legal, safety, reliability, and interoperability reasons. It involves

checking that the system adheres to technical standards, follows best practices, and meets quality benchmarks.

- **Performance:** This aspect of the validation plan focuses on how well the system performs under various conditions. It includes testing for speed, responsiveness, scalability, and efficiency. Performance validation ensures that the system meets the expected benchmarks and can handle the anticipated load without degradation in service.
- **Power Budget:** This element involves optimizing the system power consumption. It ensures the system operates within its designated power limits in different operating modes. The power budget takes into account various power states, thermal management, and the impact of different feature on power usage.
- **Interoperability:** Interoperability testing ensures that the system can work seamlessly with other systems, devices, or components. It is crucial for systems that need to interact with external entities or integrate with different technologies. This element of the validation plan checks for compatibility, data exchange, and communication protocols.
- **Security:** Security validation is essential to protect the system from threats, vulnerabilities, and attacks. It involves testing for data protection, access control, encryption, and other security measures. The goal is to ensure that the system is resilient against unauthorized access, data breaches, and other security risks.
- **Reliability, Availability, Serviceability (RAS):** This element focuses on the system's ability to perform consistently over time, remain available when needed, and be easily serviced or maintained. Reliability testing checks for stability and fault tolerance, availability testing ensures uptime and redundancy, and serviceability assesses the ease of maintenance and repair.

8.2 Some Examples

By way of example, let us suppose we are developing an edge accelerator used for quality control in a 5G-enabled smart factory. The device comes in multiple form factors, including PCIe CEM (Peripheral Component Interconnect Express Card Electromechnical Specification) and EDSFF (Enterprise and Data Center Standard Form Factor), and supports multiple distributions of the Linux operating system. Using proprietary AI models, the device monitors real-time sensor data from a network of hundreds of thousands of sensors across the factory and detects anomalies so they can be corrected as quickly as possible. For example, by combining images from 3-dimensional automated optical inspection (3D-AOI) and thermal profiles from a surface mount technology (SMT) machine, the device may predict that a piece of equipment needs to be recalibrated. By catching the issue early with predictive analytics, the device helps the factory owner prevent quality problems rather than dealing with the increased cost of yield problems on the line.

The breadth of coverage is achieved by having sufficiently capable validation software and a range of environments and configurations to exercise all of the functions of the product [19]. Register Transfer Level (RTL) design should be thoroughly verified at the unit level, cluster level, and full-chip level to ensure proper operation across all stimuli. As the model size increases, the simulation time also increases and, therefore, becomes more costly. There are diminishing marginal returns to additional simulations as the scope of coverage and model size grow very large. Consequently, it is important to thoughtfully plan this phase of work and ensure coverage meets the objectives while also satisfying the schedule constraints, as well as target certain types of coverage to emulation or post-silicon validation environments.

In this example, to ensure the validation plan has an appropriate breadth of coverage, we must include system-level tests in the various supported operating systems, include coverage with both device form factors, and ensure the combined sum of all system-level validation tests completely covers the scope of features in the MRD.

Depth of coverage refers to the extent to which a given test exercises and stresses the underlying logic. For example, suppose that our predictive analytics device can support real-time analysis of up to four automatic optical inspection (3D-AOI) images and that there are subtle differences in the way the device behaves depending on whether it is simultaneously monitoring one, two, three, or four images. Suppose the device contains two parallel sets of resources for inference. Supposing each can support two inference operations in parallel, but only one shared instance of the result handling logic, which behaves differently depending on the number of images being processed. Also, suppose that the logic handles back-to-back operations slightly differently due to initialization and clean-up of the shared resources.

If the test only covers the boundary conditions of one single image and four simultaneous images, then it could fail to exercise the logic that handles the special cases associated with processing two or three images and specific back-to-back transitions. To achieve proper depth of coverage, the validation planner must have a sufficiently deep understanding of the microarchitecture to understand the impact of different test scenarios at the logic level and plan accordingly. A close partnership with the logic design team to review the validation strategy, plans, and results throughout the program helps to ensure proper coverage. However, even with this extra attention to detail, it is possible to miss important scenarios, so additional techniques can be employed to mitigate the risk.

One common approach to ensuring proper depth of coverage is the use of pseudorandom test generators. In the example described above, an exerciser program which sends a randomly generated sequence of 3D-AOI images for inference could be developed. By seeding the random number generator and then generating test patterns deterministically from the stream of random numbers, the test can achieve the benefits of both randomness and repeatability.

During normal execution of the test, the random number generator is periodically re-seeded with a new number, and the seed is logged. Then, if a failure occurs, the test can be rerun using the failing seed. If care is taken to ensure that the test is generated deterministically from the random number sequence, then the test failure should be repeatable in a shorter time by restarting with the failing seed. This approach works well for finding logic bugs and is a good way to achieve sufficient depth of coverage across the breadth of the test suite and configurations. Of course, the results may vary depending on the nature of the issue at hand; a subtle timing bug, circuit marginality, or power integrity issue, e.g., may not be so repeatable. Even in these cases, however, the pseudorandom test generation approach arms the engineering team with the tools to help characterize the failure and understand its repeatability under varying stimuli, which are important elements of the debug process.

The topic of concurrency is another important aspect of ensuring sufficient breadth and depth of validation coverage. It is often the case in complex computer systems that a set of conceptually independent events complete successfully when they occur independently. However, they encounter problems when they occur simultaneously.

For example, consider a power management feature which detects that a given portion of the logic is idle and in order to save power, autonomously gates clocks and power to the unit. When the system subsequently needs to use the unit again, the power management feature returns the power and clocks to the unit and re-initializes it for normal operation. In systems that implement such features, it is common to implement a back pressure mechanism to notify upstream logic that the unit is temporarily unavailable. The upstream logic may have a bug that can only be exercised when the back pressure mechanism is in use.

For example, suppose we have a dispatch unit which feeds 3D-AOI images to the inference unit from a buffer in an SRAM. If the inference unit is always available, and the buffer in SRAM is sufficiently large, then the dispatch unit may never fill its buffer under normal operation. However, if the back pressure mechanism is asserted frequently or for an extended period of time, then the buffer may fill up completely (Figure 8.2).

In this sense, the inference unit power management feature may be considered as an "aggressor" or "stressor" for the dispatch unit. This buffer full scenario could exercise special logic designed to handle boundary conditions. If there is a bug in this portion of the logic, it could go undetected unless care is taken to concurrently exercise the power management features with the inference logic.

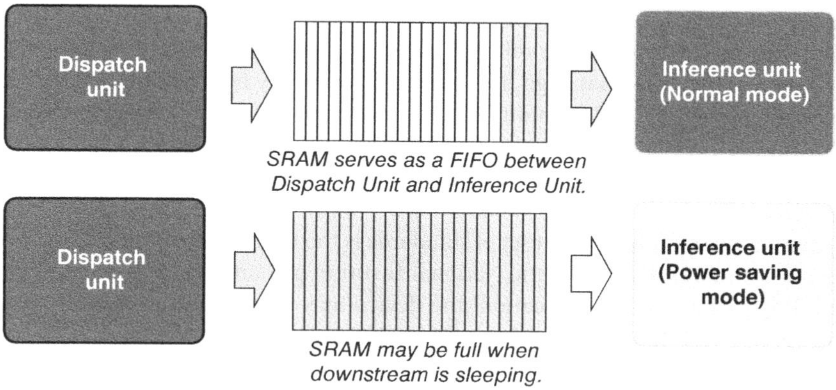

Figure 8.2 The power management feature in the downstream inference unit can be a "stressor" for the upstream dispatch unit.

This general category of validation is commonly referred to as "cross-product" coverage since proper coverage may require the product of many such "aggressor" conditions happening concurrently to find all the bugs. As a general rule, the validation planner should assume that any and all possible conditions can occur in the field after production and proactively employ pseudorandom test generation with concurrency and cross-product coverage as a means to cover future eventualities.

8.3 Standards Compliance

Standards compliance is a broad topic that will not be covered in depth in this book. However, we will discuss it briefly in the context of overall validation planning. In the 5G edge acceleration market, devices may be subject to a variety of standards from different industries. For example, with virtualized servers replacing purpose-built telecommunications equipment throughout much of the 5G network from vRAN to the 5G core, industry-standard specifications traditionally applied in server systems must be considered.

Examples include:

- JEDEC standards for memory such as DDR5 [20].
- PCIe SIG standards for PCI Express devices [21].
- Open Compute Platform standards such as DC-SCM or OAM for modular computing architecture [22, 23].

Standards which were more traditionally associated with telecommunications equipment may also apply. For example, NEBS level 3 compliance, including GR63 and GR1089 certifications, may be crucial for entering the target market. The GR63 standard [24] aims to ensure reliable operation across the more extreme environmental conditions that edge networking devices may be subject to, including broader temperature and humidity ranges, seismic activity (shock and vibration), altitude, fire resistance, and more. GR1089 [25] deals with electrical safety, electromagnetic interference (EMI) and immunity, electro-static discharge (ESD), and other areas.

Another potentially relevant set of standards is the US Federal Information Processing Standards defined by the National Institute of Standards and Technology [26]. Depending on the application, the device may be required to comply with some or all FIPS standards to gain access to the target market.

From the standpoint of validation planning, it is important to fully comprehend the requirements for adhering to each standard and the process for demonstrating compliance and achieving certification. For example, a PCI Express device must meet several functional and electrical requirements in order to comply with the specification. Specific compliance areas include electrical, configuration, link-level protocol, transaction-level protocol, and platform BIOS testing.

Demonstrating electrical compliance, e.g., includes specific lab tests to ensure transmitters and receivers meet certain bit error rate requirements when subjected to specific test patterns. A test suite is provided by the standards body, the PCI SIG (Special Interest Group), for verifying protocol compliance. There is a robust ecosystem of equipment vendors and labs that provide compliance-related equipment and services. Similar ecosystems of vendors and certification labs tend to be in place for all major standards, providing services to help system, device, and silicon vendors navigate the process of achieving compliance and gaining certifications.

8.4 Other Design Metrics

Performance is a broad topic and should be considered from multiple angles in order to achieve the desired results. At the highest level, we have end-user workload performance, which refers to how the product will perform when it is ultimately deployed and running the target workload. While this is, in a sense, the truest measure of the product's performance, it is fraught with many challenges. First, products are typically targeted at many such workloads and hence must be evaluated across the range, perhaps with some relative weighting applied. Second, the workloads may be proprietary, unavailable, prohibitively complex, or expensive to use during validation. Third, a performance issue observed with the end-user environment and workload may be unrelated to the product itself. The end user may have misconfigured their environment or implemented proprietary code poorly, thereby inadvertently degrading the performance of the product.

For the reasons described above, performance should be characterized, validated, and tuned across a variety of levels to ensure the fundamentals are correct rather than relying solely on end-user workload performance. Industry-standard benchmarks play a vital role in addressing the challenges discussed here. By standardizing the test configuration, input data, measurement methodology, and reporting, benchmarks provide a fairly consistent means for grading the performance of a given class of workload. Various vertical industries and use cases tend to have popular benchmarks that are commonly accepted and utilized for making purchasing decisions or creating marketing materials. Consequently, the process of selecting the target benchmarks and configurations to be evaluated during validation is a critical element of delivering a successful product. It is also important for validation planners to align closely with product management and/or marketing counterparts to ensure the correct benchmarks are selected and prioritized according to business needs and to set appropriate targets for each benchmark in support of the marketing objectives.

Another critical component of performance validation and tuning involves the measurement and analysis of lower-level performance primitives and/or microbenchmarks. Examples include memory and I/O access latency, throughput, cache hit/miss rates, entry and exit latencies for power management states, and other elements which play a significant role in the performance of a higher-level operation or workload. For example, suppose we are characterizing the performance of our 3D-AOI AI inference engine. Fundamentally, we would define targets in terms of latency, throughput, and scaling at the inference level. For example:

- How long does it take to complete a single inference operation from idle?
- What are the peak and sustained throughput when processing a constant stream of images?
- How does the throughput scale as we increase the number of parallel inference operations?

As we discussed above, this hypothetical inference engine can support up to four simultaneous inference operations. Throughput is often impacted by overhead as the number of parallel operations increases, so we would want to evaluate the performance as we increase the input stream from one to two to three to four parallel images.

When considering the real-world implementation of such a device, we must consider the entire software and firmware stack in addition to the hardware in order to fully comprehend the performance. As we move from a highly controlled experiment to a real-world usage model, each component of the stack introduces different complicating factors and potential problems. A typical host software stack consists of applications, middleware and libraries, device drivers, and the operating system kernel. The device itself may contain a microcontroller

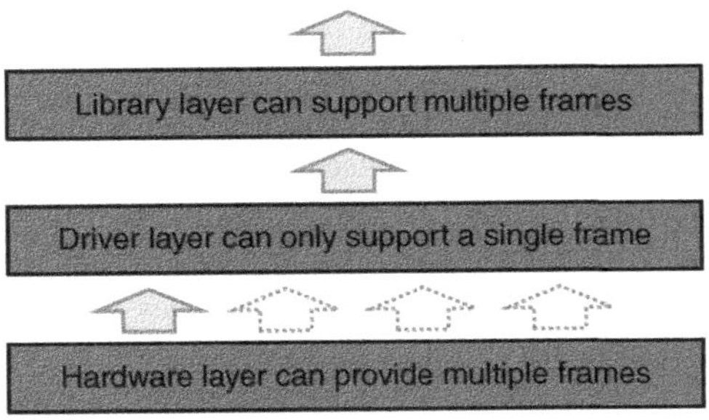

Figure 8.3 The driver implementation is the performance bottleneck in this figure. Each stack level must be optimized in order to achieve best performance.

and a firmware payload consisting of an embedded real-time operating system (RTOS), a set of embedded applications, and an FPGA bitstream to implement the programmable logic. Top performance can only be realized if each level of the stack is designed and implemented efficiently and in such a way that makes optimal use of the underlying hardware. Suppose that the library supports parallel inference of multiple frames but the driver implementation serializes the frame processing. In this case, the impact on parallel inference performance will be severe and obvious – instead of processing four frames at the same time as one, increasing throughput by approximately a factor of four, we would see no benefit when enabling parallel (batch mode) inference versus single-shot (Figure 8.3).

Other performance issues may be far more subtle. Take tail latency, for instance. We will observe a distribution if we perform the same operation repeatedly and measure the latency. It is common to observe a left-skewed distribution in latency measurements as there is usually a clear lower bound in effect as a property of the overall system, with a longer tail on the right side, indicative of the various factors that impede performance on an intermittent basis. For example, the worst-case tail latency may occur when simultaneously exiting a low power state, servicing a pending interrupt, correcting an error through a transaction replay, and encountering a high cache miss rate during the operation.

Depending on the higher-level use case, this seemingly rare confluence of events, which could be dismissed as unlikely to be worth fixing could in fact, manifest as a severe performance problem in the customer's real-world use case. Google published a noteworthy paper on this topic, which highlighted the negative impact of tail latency in disk drive accesses on interactive services at a large scale [27]. It is therefore highly recommended to include thorough latency analysis, including both typical and tail scenarios, as a key component of performance validation plans.

8.5 Summary

In this chapter, we delved into the critical engineering aspects necessary for transitioning a product from concept to production. We began with the importance of a well-defined system design process, underscored by a comprehensive MRD and Product Feature Specification. These documents are pivotal in outlining the product's intended features, compliance with industry specifications, and performance and power consumption targets.

We discussed the pivotal roles of verification, validation, and testing in ensuring that the final product aligns with the specifications set forth in the MRD. Verification, often carried out through simulation, checks the design's correctness in the pre-silicon phase. Validation, a post-silicon discipline, ensures that the product meets its specifications in real-world scenarios. Testing, particularly manufacturing test, is crucial for verifying the functionality of each unit before it reaches the customer.

The chapter highlighted the necessity of a thorough validation plan encompassing breadth and depth of coverage, standards compliance, performance, power budget, interoperability, security, and RAS. These elements are crucial for identifying potential issues, ensuring product robustness, and meeting industry standards and regulations. Through examples, we discussed the practical application of these principles, and the challenges of achieving sufficient validation coverage, the role of pseudorandom test generators, and the importance of considering concurrency in validation.

In summary, transitioning a product from concept to production is a complex and multifaceted process that requires careful planning, rigorous validation, and adherence to industry standards. By following a systematic approach and addressing the key elements discussed in this chapter, engineers can navigate this process successfully and bring innovative, reliable, and high-performing products to market.

References

1 Goodenough, J. and Aitken, R. (2010). Post-silicon is too late avoiding the $50 million paperweight starts with validated designs. *Proceedings of the 47th Design Automation Conference*, 13–18 June 2010, Anaheim California USA, pp. 8–11.

2 Mitra, S., Seshia, S.A., and Nicolici, N. (2010). Post-silicon validation opportunities, challenges and recent advances. *Proceedings of the 47th Design Automation Conference*, 13–18 June 2010, Anaheim California USA, pp. 12–17.

3 Ma, H., Panoff, M., He, J. et al. (2023). EMSim: a fast layout level electromagnetic emanation simulation framework for high accuracy pre-silicon verification. *IEEE Transactions on Information Forensics and Security* 18: 1365–1379.

4 Rusu, A., David, E., Ţopa, M. et al. (2023). Improvement and performance evaluation of an adaptive method for integrated circuits pre-silicon verification. *2023 International Symposium on Signals, Circuits and Systems (ISSCS)*, 13–14 July 2023, Iasi Romania, pp. 1–4. IEEE.

5 Mishra, P., Morad, R., Ziv, A., and Ray, S. (2017). Post-silicon validation in the SoC era: a tutorial introduction. *IEEE Design & Test* 34 (3): 68–92.

6 Manikandan, P., Mala, J., and Balamurugan, S. (2013). Post silicon functional validation from an industrial perspective. *Middle-East Journal of Scientific Research* 15 (11): 1570–1574.

7 Pointner, S. and Wille, R. (2019). Did we test enough? functional coverage for post-silicon validation. *2019 IEEE International Test Conference in Asia (ITC-Asia)*, 9–15 November 2019, Washington DC USA, pp. 31–36. IEEE.

8 Viveros-Wacher, A., Baca-Baylón, R., Rangel-Patiño, F.E. et al. (2021). Fast jitter tolerance testing for high-speed serial links in post-silicon validation. *IEEE Transactions on Electromagnetic Compatibility* 64 (2): 516–523.

9 Choi, I., Jung, W., Oh, H., and Kang, S. (2018). A debug scheme to improve the error identification in post-silicon validation. *PLoS One* 13 (9): e0202216.

10 Pandey, A., Tully, B., and Natarajan, K. (2022). High speed IO access for test forms the foundation for silicon lifecycle management. *2022 IEEE International Test Conference (ITC)*, 23–30 September 2022, Anaheim California USA, pp. 656–660. IEEE.

11 Abramovici, M., Bradley, P., Dwarakanath, K. et al. (2006). A reconfigurable design-for-debug infrastructure for SoCs. *Proceedings of the 43rd annual Design Automation Conference*, 24–28 July 2006, San Francisco California USA, pp. 7–12.

12 Bertacco, V. (2010). Post-silicon debugging for multi-core designs. *2010 15th Asia and South Pacific Design Automation Conference (ASP-DAC)*, 18–21 January 2010, Taipei, pp. 255–258. IEEE.

13 Basak, A., Bhunia, S., and Ray, S. (2016). Exploiting design-for-debug for flexible SoC security architecture. *Proceedings of the 53rd Annual Design Automation Conference*, 5–9 June 2016, Austin Texas USA, pp. 1–6.

14 Ray, S., Yang, J., Basak, A., and Bhunia, S. (2015). Correctness and security at odds: post-silicon validation of modern SoC designs. *Proceedings of the 52nd Annual Design Automation Conference*, 7–11 June 2015, San Francisco California USA, pp. 1–6.

15 Seligman, E., Haren, K., Dreyer, C., and Nayyar, R. (2008). Zero Escape plans: tying together design, simulation, and formal methods for bulletproof stepping validation. *DVCON USA*, February.

16 Chang, K.H., Markov, I.L., and Bertacco, V. (2007). Automating post-silicon debugging and repair. *2007 IEEE/ACM International Conference on Computer-Aided Design*, 4–8 November 2007, San Jose California USA, pp. 91–98. IEEE.

17 Cao, Y., Zheng, H., and Ray, S. (2019). A communication-centric observability selection for post-silicon system-on-chip integration debug. *20th International Symposium on Quality Electronic Design (ISQED)*, 6–7 March 2019, Santa Clara California USA, pp. 278–283. IEEE.

18 Intel Corporation Annual Report 1994. https://www.intel.com/content/www/us/en/history/history-1994-annual-report.html, Intel Corporation, 1994 (accessed 17 March 2024).

19 Mehta, A.B. (2018). *ASIC/SoC Functional Design Verification. A Comprehensive Guide To Technologies and Methodologies*. Springer.

20 DDR5 SDRAM Standard, JESD79-5C.01, Joint Electronic Engineering Council (JEDEC), July 2024. https://www.jedec.org/standards-documents/docs/jesd79-5b. (accessed 1 September 2024).

21 Peripheral Component Interconnect Special Interest Group Standards (PCI SIG Standards) Home Page. https://pcisig.com/ (assessed 1 September 2024).

22 Raghu P., Shaw M., Chauhan P., Tavallaei S., Branch M., Possing M. Open Compute Project Datacenter Secure Control Module Specification. Revision 1.0, 11 March 2021. https://www.opencompute.org/documents/ocp-dc-scm-spec-rev-1-0-pdf (accessed 19 March 2024).

23 OCP OAI Workstreams. Open Compute Project Standard OCP Accelerator Module (OAM) Specification v1.5, 18 October 2021. https://www.opencompute.org/documents/ocp-accelerator-module-design-specification-v1p5-final-20220223-docx-1-pdf (accessed 19 March 2024).

24 NEB Requirements: Physical Protection, Telcordia, Document Number GR63, December 2017. https://telecom-info.njdepot.ericsson.net/site-cgi/ido/docs.cgi?ID=SEARCH&DOCUMENT=GR-63 (accessed 19 March 2024).

25 Telcordia Electromagnetic Compatiblity (EMC) and Electrical Safety - Generic Criteria for Network Telecommunications Equipment, Telcordia, Document Number GR1089, July 2022. https://telecom-info.njdepot.ericsson.net/site-cgi/ido/docs.cgi?ID=SEARCH&DOCUMENT=GR-1089 (accessed 19 March 2024).

26 Federal Information Processing Standards (FIPS) Home Page. National Institute of Standards and Technology (NIST). https://www.nist.gov/standardsgov/compliance-faqs-federal-information-processing-standards-fips (accessed 1 September 2024).

27 Barroso, L.A. (2011). Warehouse-scale computing: entering the teenage decade. *ISCA Keynote, SJ* 12: 25–60.

Part V

Future Roadmap

9

The Road Ahead

In the post-Moore Law era, edge computing acceleration technology is not only guided by the underlying process technology but also driven by the underlying computing technology and the prevailing applications.

Over the past four decades, we have witnessed a remarkable evolution in computing, transitioning from mainframe computing to PC computing, then to Internet computing, and now to 5G mobile computing. The upcoming mobile computing may set to be spatial computing, which promises to spawn numerous opportunities. The total spatial computing market is projected to exceed $1 trillion in size in the next decade.

Spatial computing is revolutionizing how we interact with the digital world by integrating it seamlessly into our physical surroundings [1]. This technology merges the realms of artificial intelligence (AI), augmented reality (AR), and virtual reality (VR), creating a cohesive and interactive experience that transcends conventional computing applications. At the heart of spatial computing lies AI, which empowers the system to understand and analyze the environment and create appropriate content. This intelligent processing forms the foundation of spatial computing, enabling dynamic and context-aware interactions.

The recently released Apple Vision Pro stands as the world's first multitasking spatial computer, heralding a new era in mixed reality (MR) technology and using AI to merge VR with AR. It creates an immersive and interactive experience that blurs the lines between the digital and physical worlds. Users can seamlessly run multiple MR applications where digital information is superimposed onto real-world surroundings.

To capture the full potential of spatial computing, the edge computing acceleration architecture will require disruptive edge acceleration technology in the road ahead. Several key factors influence and shape the architecture:

Ultra-Energy Efficiency: As computing becomes increasingly pervasive in all aspects of life, there is a growing demand for ultra-energy efficiency [2]. Portable

Edge Computing Acceleration: From 5G to 6G and Beyond, First Edition.
Patrick Hung, Hongwei Kan, and Greg Knopf.

spatial computers need to be both powerful and energy efficient. As a result, it makes sense to offload much of the processing to the edge, ensuring that portable spatial computers are small and ultra-energy efficient.

Artificial Intelligence: Generative and discriminative AI have gained significant momentum in recent years, driving the need for specialized computing hardware and software architecture to support complex AI algorithms. The latest computing technology trends focus on AI-oriented processors, neural network accelerators, and ML frameworks to enable advanced AI applications [3, 4]. Many simple discriminative AI tasks require very low latency and can be integrated inside each portable spatial computer. Complicated generative AI tasks can be offloaded to the edge.

Emerging IoT Infrastructure: Hundreds of IoT devices will need to work seamlessly in the spatial computing environment. The proliferation of IoT devices and applications drives the need for computing technologies that can seamlessly connect, process, and analyze data from diverse IoT endpoints. Edge computing, fog computing, and IoT-optimized processors are some of the trends that cater to the unique requirements of IoT deployments [2, 5].

Technological Convergence: The convergence of technologies, such as computing, wireless networking, AI, and IoT, drives the latest trends. Interdisciplinary research and integration of technologies enable innovative solutions and applications [6].

User Experience and Interface Innovation: The latest trends in computing technology are influenced by the desire to improve user experience in Metaverse and Omniverse through innovative AR and VR interfaces [7].

Edge and Cloud Synergy: Edge computing complements cloud computing by enabling data processing closer to the data source, supporting microsecond latency and 100 Mbp/s bandwidth. The latest trends aim to strike a balance between edge and cloud computing, creating a seamless and integrated computing ecosystem [7].

Demand for Security and Privacy: With the increasing digitization of information and cyber threats, there is a strong emphasis on security and privacy in this post-quantum computing era. Research focuses on developing post-quantum secure hardware, encryption techniques, and privacy-preserving computing solutions [2, 5].

Standardization and Green Initiatives: International collaboration and standardization efforts play a critical role in shaping the latest trends in computing technology. Organizations like IEEE, ISO, and ITU work toward establishing global standards and best practices [5]. The computing industry is increasingly conscious of its environmental impact. Green computing trends focus on developing solar-operated energy-efficient hardware, achieving carbon-neutral sustainable practices in computing [8].

Other Acceleration Technologies: Quantum computing promises to solve complex intractable problems exponentially faster than classical computers. Unfortunately, there is no general-purpose quantum computer at this moment. Research in general-purpose quantum computing is driving advancements in quantum algorithms, quantum hardware, and quantum programming languages [8]. Unfortunately, general-purpose quantum computing may be decades away.

As computing becomes more integral to daily life, trends are shaped by the need to make technology affordable and accessible to a broader population. This includes the development of budget-friendly devices, low-cost cloud services, and inclusive computing solutions [8]. The latest trends in computing technology are driven by a complex interplay of technological advancements, market forces, societal demands, and the pursuit of transformative solutions. These trends call for technologies to handle massive amounts of data, high-speed and ultra-low latency communications. They pave the way for an increasingly connected, intelligent, and efficient future.

9.1 Spatial Computing and Networking

The convergence of computing and wireless networking in spatial computing refers to the integration and seamless interaction between computing technologies and wireless communication infrastructure. It represents the coming together of two essential components of modern technology to create a unified and interconnected ecosystem.

In the context of this convergence, we note the following trends:

Computing Technologies: Computing technologies include hardware, software, and data processing capabilities. This encompasses a wide range of devices, from traditional personal computers and servers to smartphones, tablets, IoT devices, and even edge computing devices. Computing technologies handle data processing, storage, and execution of various applications and services.

Wireless Networking: Wireless networking refers to the technology that allows devices to communicate and exchange data without the need for physical wired connections. It includes various wireless communication standards such as Wi-Fi, Bluetooth, Zigbee, cellular networks (e.g., 5G and 6G), and other wireless protocols used in IoT devices.

The convergence of computing and wireless networking leads to a synergetic relationship, where computing devices leverage wireless connectivity to access cloud services, share data, and interact seamlessly with other devices and users. Here are some key aspects of the convergence:

Mobility and Connectivity: The convergence enables users to access computing resources and services virtually anywhere using wireless networks. Smartphones, laptops, and IoT devices can connect to the internet and cloud services on the go, providing constant connectivity and mobility [9].

Cloud Computing: Cloud computing plays a vital role in the convergence. By utilizing wireless networks, computing devices can access cloud services and applications, offloading complex computations, and data storage to remote data centers [9].

Internet of Things (IoT): The integration of computing and wireless networking is a cornerstone of IoT. IoT devices communicate with each other and with cloud-based applications over wireless networks, creating an interconnected network of smart devices [10].

Edge Computing: The convergence of computing and wireless networking extends to the edge of the network, where edge computing devices process data locally, reducing latency and leveraging wireless connectivity to interact with the cloud and other devices [9].

Wireless Sensor Networks: The convergence enables the deployment of wireless sensor networks, where sensor nodes collect data and transmit it wirelessly to computing devices for analysis and decision-making [11].

Mobile Applications and Services: Mobile applications and services rely on the convergence of computing and wireless networking to provide real-time interactions, location-based services, and seamless user experiences [9].

5G and Beyond: The evolution of wireless networking, particularly with the deployment of 5G and beyond, further enhances convergence by offering higher data speeds, lower latency, and increased capacity, supporting more advanced computing applications and services [9].

The convergence of computing and wireless networking has revolutionized the way we access information, communicate, and interact with technology. It has led to the proliferation of mobile computing, IoT devices, edge computing, and cloud services, enabling a connected world with seamless data exchange and ubiquitous access to computing resources.

9.2 Supporting 5G/6G Spatial Computing and Networking

5G and future 6G technologies are poised to play a crucial role in further supporting and enhancing the convergence of computing and wireless networking.

Specifically, much of the processing can be performed on the edge accelerators, thanks to the ultra-low latencies. This is especially important in robotic and spatial computing applications.

Faster Data Speeds and Lower Latency: 5G and future 6G networks offer significantly faster data speeds and lower latency compared to their predecessors. This enables quicker data transfers and response times, making real-time computing and communication more seamless [12]. As a result, applications that rely on instant data processing, such as edge computing, AR, VR, and autonomous systems, can operate more effectively.

Edge Computing Enablement: The low latency and high bandwidth capabilities of 5G and future 6G networks enable efficient support for edge computing. By processing data closer to the edge of the network, computing devices can offload tasks from central cloud servers, reducing latency and ensuring more rapid decision-making. This is especially critical for applications that require real-time processing and decision-making, such as autonomous vehicles, smart city infrastructure, and industrial automation [8, 12].

Massive Device Connectivity: Both 5G and future 6G are designed to support a massive number of connected devices simultaneously. This capacity is crucial for the growth of the IoT and the deployment of various smart devices. The convergence of computing and wireless networking will heavily rely on the ability to handle the vast number of connected devices generating data and communicating with each other and cloud services [12].

Network Slicing and Customization: Network slicing is a key feature of 5G that allows the creation of virtual networks optimized for specific use cases. Future 6G networks are likely to build on this capability. Network slicing enables customized network configurations for different applications, industries, and services, ensuring efficient use of resources and meeting diverse computing and communication requirements [13].

Enhanced Security and Privacy: Both 5G and future 6G technologies are expected to enhance security and privacy measures. As the convergence of computing and wireless networking leads to an increasingly interconnected ecosystem, robust security mechanisms will be essential to protect data and devices from cyber threats. Advancements in encryption, authentication, and privacy-preserving techniques will be critical in ensuring secure and trustworthy communication [8, 13].

Integration of AI and Machine Learning: 5G and future 6G networks are likely to integrate AI and machine learning algorithms to optimize network operations, improve resource allocation, and enhance overall efficiency. AI-driven wireless networking will play a vital role in managing complex networks with a vast number of connected devices and services, enabling intelligent and autonomous decision-making in real-time [12].

Enabling New Applications and Use Cases: The high-speed, low-latency, and massive connectivity capabilities of 5G and future 6G open up new possibilities for applications and use cases that leverage the convergence of computing and wireless networking. These technologies can support advancements in areas like smart infrastructure, remote healthcare, connected vehicles, immersive AR/VR experiences, and smart manufacturing, among others [12].

5G and future 6G technologies form a solid foundation for the continued convergence of computing and wireless networking. Their fast data speeds, low latency, edge computing support, massive device connectivity, network slicing, security enhancements, AI integration, and enabling new applications will pave the way for even more innovative and interconnected technologies in the future. The synergy between these advanced wireless networks and computing technologies will drive the evolution of our connected world, providing unprecedented opportunities for efficiency, convenience, and transformative applications across various industries.

9.3 Migrating to 6G

Work on 6G started in 2018 when the 3GPP 5G standard (Release 15) was published. An ITU focus group has published Network 2030, which covers a blueprint of technology, applications, and market drivers toward the year 2030 and beyond. At the time of this writing, the transition from 5G to 6G is still in its early stages, and concrete specifications for 6G have not been finalized. However, there are ongoing research and development efforts, discussions, and proposals to pave the way for the future of 6G communications. In addition to key topics of higher frequencies and data rates, extreme capacities and connectivity, internetworking and AI, which we have discussed in previous sections, here are other key aspects and considerations related to the transition from 5G to 6G:

Research and Standardization: The development of 6G is driven by various research institutions, technology companies, and standardization bodies. Researchers are exploring potential technologies and concepts that could shape 6G networks, such as terahertz frequencies, new waveforms, advanced antenna technologies, and novel communication architectures [8].

Integrated Satellite and Terrestrial Networks: 6G is exploring the integration of satellite and terrestrial networks to extend coverage and improve global connectivity. This integration could lead to seamless communication experiences, especially in remote and underserved areas [8, 14].

Energy Efficiency and Sustainability: As with each new generation of communication technology, energy efficiency and sustainability will be important considerations for 6G. Research is focused on developing energy-efficient components, networking protocols, and power management strategies to minimize environmental impact [8, 15].

Global Collaboration and Policy Development: The development of 6G involves international collaboration between industry players, governments, and standardization bodies. Policymakers are engaging in discussions about spectrum allocation, regulatory frameworks, and global standards to ensure harmonized deployment and interoperability of 6G networks worldwide [8].

It is important to note that 6G is still in the early research phase, and widespread commercial deployment is likely several years away. The transition from 5G to 6G will involve continuous advancements in wireless communication technologies, significant standardization efforts, and substantial investments in research and development by the telecommunications industry and other stakeholders. As research progresses and technology advances, more concrete details about 6G's capabilities and features will emerge, shaping the next generation of communication networks.

9.3.1 Cutting Edge 6G Research

The following is an overview of the general concepts and potential features that researchers and industry experts are exploring for spatial computing.

Hyper-Connected and Intelligent Networks: In 6G, the vision is to create a hyper-connected world where various wireless networks seamlessly integrate to provide ubiquitous connectivity. This integration will enable users to switch between different network types (e.g., cellular, Wi-Fi, and satellite) seamlessly, ensuring uninterrupted communication regardless of location or device [8].

AI-Driven Network Orchestration: AI will be at the core of 6G networks, driving intelligent network orchestration and optimization. Advanced machine learning algorithms will continuously analyze network data, user behavior, and traffic patterns to dynamically allocate resources, reduce latency, and predict user needs [8].

Terahertz (THz) and Beyond Frequencies: 6G aims to explore higher-frequency bands beyond the capabilities of previous generations, including terahertz (THz) frequencies. These extremely high frequencies offer vast amounts of available spectrum, enabling unprecedented data rates and capacity [8].

Massive Multiple Input Multiple Output (MIMO) and Advanced Beamforming: Building on the success of massive MIMO and beamforming in 5G, 6G will employ even more advanced antenna arrays and beamforming techniques. This will enhance spatial multiplexing, allowing multiple users to be served simultaneously with high data rates and improved coverage [8].

Tactile Internet and Ultra-low Latency: One of the ambitious goals of 6G is to achieve ultra-low latency, enabling real-time and interactive applications with haptic feedback. The tactile Internet will revolutionize fields like remote surgery, VR, and the teleoperation of robots in hazardous environments [8].

Extreme Edge Computing: Edge computing will be further advanced in 6G to reduce latency and bandwidth consumption. Edge nodes will be equipped with powerful computing resources, enabling real-time data processing and analytics closer to the end-users or devices [14].

Satellite Integration and Global Coverage: 6G may heavily integrate satellite communication to achieve global coverage, especially in remote and rural areas. Low Earth Orbit (LEO) satellite mega-constellations will play a crucial role in providing low-latency, high-bandwidth connectivity to users worldwide [14].

Green and Sustainable Networks: Sustainability will be a significant focus in 6G architecture. Energy-efficient designs, renewable energy integration, and smart power management strategies will be adopted to reduce the carbon footprint of wireless networks [8].

Network Slicing and Customized Services: 6G will extend the concept of network slicing to offer highly customized virtual networks tailored to specific applications and industries. This enables efficient resource allocation and the provision of optimized services for different use cases [13].

Enhanced Security and Privacy Measures: As data becomes more critical in 6G applications, robust security measures will be essential. 6G networks will employ advanced encryption, secure authentication protocols, and privacy-preserving technologies to safeguard user data and protect against cyber threats [8, 13].

Intelligent Mobility and Seamless Handover: 6G will emphasize intelligent mobility management, ensuring smooth and seamless handovers between different networks and technologies as users move across diverse coverage areas [8].

Beyond Communication: Converging Technologies: In addition to communication, 6G is likely to explore convergence with other emerging technologies, such as AR, AI, blockchain, and the IoT, to enable a new era of immersive experiences and innovative applications [12].

It is important to remember that 6G is still a developing concept, and the specifics of its architecture will continue to evolve through research, standardization, and industry collaborations. As the technology matures, more concrete details about the architecture and its capabilities will emerge, setting the stage for the next generation of wireless communication.

9.4 Enabling Technologies for 5G and Beyond

In previous chapters, we have already discussed many key enabling technologies for 5G, such as millimeter wave, massive MIMO, small cells and beamforming. They are expected to continue to support 6G development. In this section, we will further focus on how technologies in processing-in-memory (PIM), electronic packaging, new memory architecture, and AI may help accelerate data processing to support the realization of the upcoming spatial computers.

9.4.1 Processing-in-Memory Architecture

PIM architecture is extremely important in spatial computing. It can significantly accelerate data processing by reducing data movement and improving the overall system efficiency [16]. Traditional computing systems, including CPUs and GPUs, often suffer from a "memory wall" or "von Neumann bottleneck," where the speed of data processing is limited by the slower speed of accessing data from memory. PIM addresses this limitation by bringing computations closer to the data rather than moving data back and forth between memory and processing units. See Figure 9.1.

Here are some ways PIM architecture helps accelerate data processing:

Data Locality and Reduction in Data Movement: In traditional architectures, data processing involves moving data from the main memory (RAM) to the processor cache or registers, where computation occurs. This data movement is a time-consuming process and leads to a memory bottleneck. In PIM, computation units are integrated into the memory subsystem. This integration brings computation closer to where the data is stored, reducing the need for

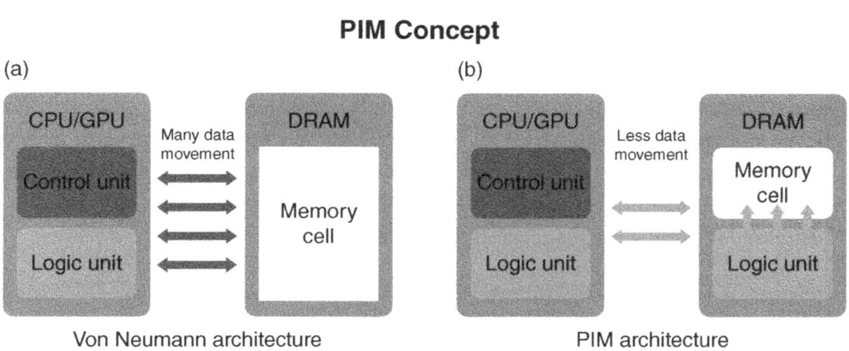

Figure 9.1 Concept of PIM. Source: https://www.eetimes.com/the-prospect-of-processing-in-memory-pim-in-memory-systems-for-ai-applications.

data movement [16–18]. Data can be accessed and processed directly within the memory, minimizing the time and energy required for data transfers.

Parallelism and Concurrency: The PIM architecture allows for exploiting parallelism and concurrency more efficiently [17]. With processing units distributed throughout the memory, multiple operations can be performed concurrently on different data elements. This parallelism enables faster processing of large datasets and complex computations, improving overall system performance.

Customization and Specialization: PIM systems can be designed to cater to specific workloads or data types. By integrating specialized processing units into memory, such as accelerators optimized for machine learning tasks or graph processing [18], PIM can significantly boost performance for these specific applications. This specialization is challenging to achieve in traditional architectures, where the same general-purpose processing units handle all tasks.

Energy Efficiency: Data movement is one of the most energy-consuming operations in modern computing systems. PIM architecture helps reduce the amount of data movement, leading to lower energy consumption [16, 18]. By performing computation in memory, PIM systems can achieve higher energy efficiency and extend the battery life in mobile devices or reduce power consumption in data centers.

Memory Bandwidth Utilization: In traditional systems, multiple processing units competing for memory access can saturate the memory bandwidth, causing contention and performance degradation. In PIM architectures, computation units can efficiently access data locally within the memory, leading to better utilization of memory bandwidth and reduced contention [18].

Scalability: PIM can enhance the scalability of computing systems, especially for memory-intensive workloads. As data processing demands grow, traditional architectures may face scalability challenges due to limited memory bandwidth and increasing data movement. PIM can help address these issues by processing data locally, improving overall system scalability [18].

Data-Intensive Applications: PIM architecture is particularly well-suited for data-intensive applications involving large datasets and frequent data access [18]. Examples of such applications include data analytics, machine learning, databases, and graph processing. PIM's ability to perform computation in proximity to data significantly accelerates these data-centric tasks.

Despite its promising advantages, PIM also faces some challenges. Designing efficient memory hierarchies, ensuring data consistency, handling errors, and maintaining compatibility with existing software and programming models are among the complex tasks that need to be addressed to fully realize the potential of PIM.

PIM remains an active area of research and development as of today. Ongoing efforts in this field aim to overcome challenges and eventually bring PIM architectures to practical use, enabling even more efficient and powerful data processing capabilities in future computing systems.

9.4.2 New Packaging Architecture

New packaging architectures, such as chiplet heterogeneous integration, can significantly accelerate data processing by improving performance, reducing power consumption, and enhancing flexibility and scalability. This is especially important in spatial computing, where many types of acceleration cores are required at the same time. To take advantage of the economy of scale, it makes sense to build SOCs, GPUs, FPGAs, memories, and sensors on separate wafers and then integrate them together using 2.5D or 3D chiplets.

Chiplets are discrete, smaller semiconductor components that can be assembled and integrated into a single package, creating a more modular and flexible approach to building electronic systems. See Figure 9.2 for an illustration.

Here is how chiplets can accelerate data processing:

Specialization and Heterogeneity: Chiplets enable the integration of specialized components optimized for specific tasks [20]. For instance, a chiplet can be designed to excel at AI computations, while another focuses on high-speed networking. This specialization allows each chiplet to perform its designated function more efficiently, leading to faster data processing for specific workloads.

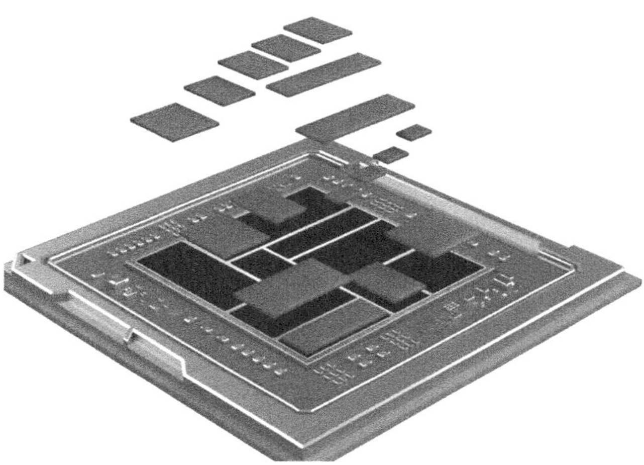

Figure 9.2 A disaggregated SoC comprising multiple chiplets [19].

Improved Yield and Cost Efficiency: As semiconductor manufacturing processes advance, producing large monolithic chips with high yields becomes increasingly challenging and expensive. Chiplets provide a solution by using smaller, more manageable components. Smaller chips are more likely to be defect-free, resulting in higher yields and lower manufacturing costs [20, 21].

Advanced Interconnects: Chiplets often use advanced packaging technologies, such as 2.5D and 3D integration, which offer shorter and faster interconnect paths between the chiplets [20]. This reduced interconnect length leads to lower latency and higher bandwidth, enhancing data transfer rates between chiplets and accelerating data processing.

Distributed Processing and Parallelism: With chiplets, the data processing workload can be distributed across multiple specialized components [22]. Each chiplet can independently execute computations in parallel, increasing the overall system throughput and reducing processing time. This parallelism is particularly beneficial for tasks that can be divided into smaller independent sub-tasks, such as data analytics and simulations.

Reduced Latency and Bandwidth Constraints: The integration of chiplets can reduce the overall system's latency [22]. Since chiplets are placed closer together, data can move faster between them, mitigating latency and bandwidth constraints that could hinder performance in traditional architectures.

Modularity and Upgradability: Chiplets enable modular system design, allowing individual components to be easily upgraded or replaced without changing the entire system [20]. This modularity not only simplifies maintenance and repairs but also facilitates future technology upgrades, extending the lifespan and performance of the system.

Scalability and Flexibility: With chiplets, system designers can tailor the number and types of chiplets based on the specific requirements of a given application [20]. This scalability allows systems to be customized for various use cases, from energy-efficient edge devices to high-performance data centers, providing flexibility to meet diverse demands.

Heterogeneous Integration: Chiplets enable the integration of different technologies within a single package [20]. This heterogeneous integration allows for combining components with various functionalities, like compute chiplets, high-bandwidth memory (HBM) chiplets, and other specialized units. As a result, data processing bottlenecks associated with memory-bound applications can be alleviated, leading to improved performance.

Power Efficiency: Chiplets can contribute to lower power consumption by enabling more efficient power management [20, 22]. As each chiplet can operate independently, the system can direct power only to the chiplets required for a particular workload, while others can remain in a low-power state. This fine-grained power management reduces overall power consumption and increases energy efficiency.

Faster Time to Market: Chiplet-based designs facilitate faster product development cycles [20]. By reusing validated and pre-qualified chiplets, system designers can focus on integrating and optimizing these components rather than starting from scratch. This can accelerate the time to market for new products, allowing companies to stay competitive and innovate rapidly.

Diverse Applications: The flexibility and modularity of chiplets make them suitable for a wide range of applications, from consumer electronics and mobile devices to high-performance computing and data centers. Each application can benefit from chiplet-based designs tailored to its specific requirements, leading to optimized performance and efficiency [21].

As chiplet technologies continue to evolve and gain wider adoption, they have the potential to revolutionize the semiconductor industry and the design of computing systems. By leveraging the benefits of specialization, parallelism, modularity, and heterogeneous integration, chiplets accelerate data processing, enable more efficient and powerful systems, and foster a more collaborative and innovative ecosystem among hardware vendors and system designers.

9.4.3 New Memory Architecture

New memory architectures, such as High Bandwidth Memory 3 (HBM3) and High Bandwidth Memory 4 (HBM4), can accelerate data processing in several ways by providing faster data access, higher bandwidth, lower power consumption, and improved system performance. HBM is an advanced memory technology designed to meet the increasing demands of data-intensive applications. For instance, HBM3 offers a peak data transfer rate of 6.4 Gbps [23], while HBM4 is expected to double that.

A HBM structure is illustrated in Figure 9.3.

Here is how HBM can accelerate data processing:

3D-Stacked Memory Design: HBM utilizes a 3D stacking approach where memory dies are vertically stacked on top of each other. This design allows for a large amount of memory to be packed in a small physical footprint, reducing the distance between memory cells and improving data transfer rates. The 3D stacking also results in shorter interconnects, which reduces signal propagation delays and latency [25].

Wide Data Bus and High Bandwidth: HBM features a wide data bus that allows for the simultaneous transfer of multiple data chunks in a single cycle. This wider bus enables higher data transfer rates and significantly increases memory bandwidth compared to conventional memory technologies like DDR4. The increased bandwidth is essential for memory-intensive tasks, such as large-scale data processing, high-resolution graphics, and AI applications [24, 25].

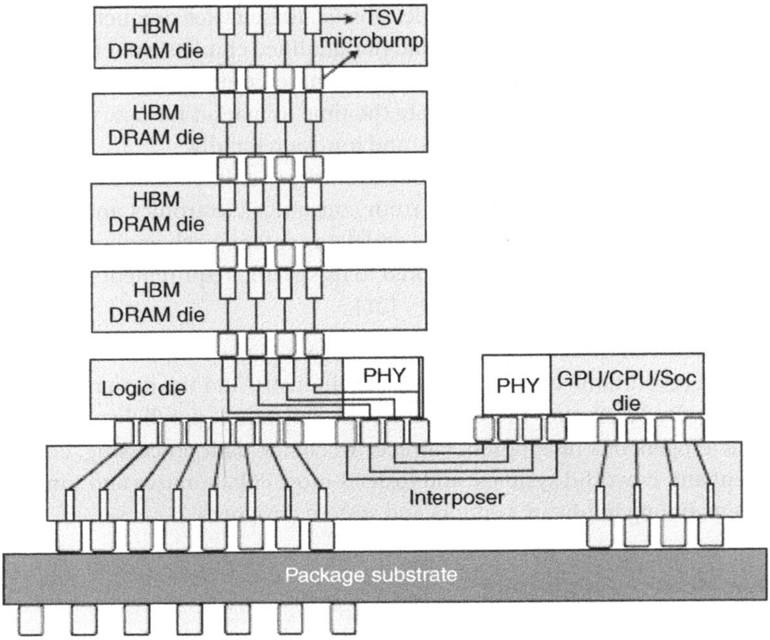

Figure 9.3 HBM structure showing memory chips arranged in a 3D-stacked format [24]. Shantharama, P et al., 2020/IEEE/CC BY 4.0.

Through-Silicon Vias (TSVs): HBM utilizes TSVs to enable communication between the vertically stacked memory dies. TSVs provide high-density vertical connections that improve data transfer efficiency and enable the high bandwidth achievable in HBM [24, 25].

Low Power Consumption: HBM is designed to be power-efficient, consuming less energy per unit of data transferred compared to traditional memory technologies. The reduced power consumption is beneficial for power-constrained devices like mobile phones, laptops, and other energy-efficient computing systems [25].

Heterogeneous Integration and Co-Packaging: HBM enables heterogeneous integration, allowing for different types of chips, such as processors, GPUs, and specialized accelerators, to be co-packaged with HBM3 memory. This integration enhances data processing performance by reducing the data movement between different chips and improving overall system efficiency [25].

High Memory Density and Capacity: HBM's 3D stacking technology enables the integration of a large memory capacity within a small physical footprint.

This high memory density is crucial for memory-intensive applications that require large working sets and extensive data storage [24, 26].

Support for High-Performance Computing and AI: HBM's high memory bandwidth and low latency make it well suited for high-performance computing (HPC) applications and AI workloads. These applications often involve complex data processing and analysis, where quick access to vast amounts of data is crucial for optimal performance. Tasks such as high-resolution graphics rendering, scientific simulations, and AI workloads can benefit greatly from HBM3's and HBM4's capabilities [26].

Scalability and Future-Proofing: HBM memory can be scaled by adding more memory stacks, allowing systems to be tailored to the specific requirements of different applications. This scalability ensures that HBM3 can accommodate future memory demands and technologies without requiring a complete redesign of the memory subsystem [27].

In summary, HBM memory architecture significantly boosts data processing performance by providing high memory bandwidth, lower latency, improved power efficiency, and large memory capacities. Its 3D-stacked design, wide data bus, and support for heterogeneous integration make it a compelling choice for data-intensive applications, high-performance computing, and AI workloads. As technology advances, HBM will play a critical role in meeting the growing demands of spatial computing systems.

9.4.4 Artificial Intelligence-Driven Architectures

New AI technologies, such as generative AI and mixed-signal AI, have the potential to accelerate data processing in various ways, primarily by enhancing efficiency, automation, and problem-solving capabilities. These AI technologies can contribute to faster data processing in the following areas:

Generative AI:

Data Augmentation: Generative AI can create synthetic data that closely resembles real data in a given domain. By augmenting existing datasets with this synthetic data, the size and diversity of the training data for machine learning models can be increased. As a result, the models can better capture the underlying patterns and variations in the data, leading to improved generalization and higher accuracy when applied to new, unseen data. Data augmentation with generative AI is widely used in image recognition, natural language processing, and other data-intensive tasks [28].

Data Completion and Imputation: Missing data is a common challenge in various domains, ranging from medical records to financial data. Generative AI can help fill in missing values in datasets, enabling more complete and usable data for analysis and modeling [29]. The ability to impute missing data is particularly valuable in situations where data collection is expensive or time-consuming.

Synthetic Data for Simulation: In fields such as robotics, autonomous vehicles, and VR, generating realistic synthetic data is crucial for training and testing models. Generative AI can produce synthetic data that mimic real-world scenarios, providing valuable datasets for simulation and experimentation. This capability accelerates research and development in these domains by reducing the reliance on costly and potentially hazardous real-world data collection [30, 31].

Data Generation for Creativity and Content Creation: Generative AI has opened up new avenues for creativity in content creation. For instance, it can be used to generate art, music, and text, empowering artists and content creators with new tools for inspiration and content generation [32]. In industries like gaming and entertainment, generative AI can automate the generation of game assets, character designs, and immersive environments, speeding up content creation and reducing manual labor [33].

Mixed-signal AI:

Analog Processing Efficiency: Mixed-signal AI leverages the strengths of both analog and digital processing. Analog computing excels at continuous signal processing, optimization, and control tasks. By utilizing analog elements for specific computations, mixed-signal AI can achieve higher computational efficiency, reducing the time required for certain operations [34].

Faster Sensor Data Processing: In IoT and sensor-driven applications, mixed-signal AI can efficiently process continuous sensor data in real-time. For instance, mixed-signal AI can rapidly analyze sensor data and make decisions based on real-time input in environmental monitoring, industrial automation, and healthcare applications. This capability is essential for quick responses to changing conditions and events [35, 36].

Energy Efficiency for Edge Devices: Edge devices, such as smartphones, wearables, and IoT devices, often have limited power resources. Mixed-signal AI can offload certain computations to analog processing elements, reducing the power consumption of the device. This optimization results in longer battery life and improved energy efficiency, making mixed-signal AI particularly valuable for low-power devices [36, 37].

Signal Processing and Noise Reduction: Mixed-signal AI is well-suited for applications that involve signal processing and noise reduction. By leveraging analog processing techniques, mixed-signal AI can effectively handle noisy sensor data and enhance signal quality. This capability is valuable in audio processing, image denoising, and communications, where signal fidelity is crucial for accurate data processing [37].

As an example application, AI can be applied to analog and mixed-signal circuit design [38]. AI can assist in automating the design process, optimizing parameters, and exploring vast design spaces to find the most efficient and reliable solutions. It can also help with analog layout generation, reducing design time and human error in the layout process. Reinforcement learning can be employed to optimize circuit performance by learning from simulations and real-world data.

In summary, generative AI and mixed-signal AI are two emerging technologies that offer unique ways to accelerate data processing in various applications. Generative AI enhances data processing by generating synthetic data for data augmentation, completion, and creative content generation. On the other hand, mixed-signal AI leverages both digital and analog processing to achieve higher computational efficiency, faster sensor data processing, and improved energy efficiency, particularly in edge devices. As these technologies continue to evolve, they are likely to play a significant role in optimizing data processing and enhancing AI-driven applications in diverse fields.

9.5 Some Final Thoughts

Spatial computing with 5G/6G edge acceleration indeed represents an exciting frontier in the world of technology. By combining advanced connectivity with AI, MR, AR, VR, global positioning systems (GPSs), and indoor positioning systems, this technology promises the creation of immersive and interactive experiences that seamlessly integrate the digital and physical worlds.

The high-speed, low-latency capabilities of 5G and the future potential of 6G are crucial for enabling real-time data processing and transmission, essential for the smooth operation of spatial computing applications. These applications can range from interactive MR/AR/VR experiences, digital twins, precision navigation to smart cities and industrial automation.

As the technology evolves, we can expect to see a proliferation of edge acceleration technologies designed to optimize the processing of spatial data. These might include specialized hardware accelerators, edge computing platforms, and advanced networking solutions that ensure data is processed as close to the source as possible, reducing latency and improving efficiency.

References

1 Shekhar, S., Feiner, S.K., and Aref, W.G. (2015). Spatial computing. *Communications of the ACM* 59 (1): 72–81.

2 Narayanan, A., De Sena, A.S., Gutierrez-Rojas, D. et al. (2020). Key advances in pervasive edge computing for industrial internet of things in 5g and beyond. *IEEE Access* 8: 206734–206754.

3 Riggio, R., Coronado, E., Linder, N. et al. (2021). AI@EDGE: a secure and reusable artificial intelligence platform for edge computing. *2021 Joint European Conference on Networks and Communications & 6G Summit (EuCNC/6G Summit)*, Virtual Conference, 8–11 June 2021, pp. 610–615. IEEE.

4 Letaief, K.B., Shi, Y., Lu, J., and Lu, J. (2021). Edge artificial intelligence for 6G: vision, enabling technologies, and applications. *IEEE Journal on Selected Areas in Communications* 40 (1): 5–36.

5 Sahmim, S. and Gharsellaoui, H. (2017). Privacy and security in internet-based computing: cloud computing, internet of things, cloud of things: a review. *Procedia Computer Science* 112: 1516–1522.

6 De Lima, C., Belot, D., Berkvens, R. et al. (2021). Convergent communication, sensing and localization in 6G systems: an overview of technologies, opportunities and challenges. *IEEE Access* 9: 26902–26925.

7 Buyya, R., Srirama, S.N., Casale, G. et al. (2018). A manifesto for future generation cloud computing: research directions for the next decade. *ACM Computing Surveys (CSUR)* 51 (5): 1–38.

8 De Alwis, C., Kalla, A., Pham, Q.V. et al. (2021). Survey on 6G frontiers: trends, applications, requirements, technologies and future research. *IEEE Open Journal of the Communications Society* 2: 836–886.

9 Tang, X., Cao, C., Wang, Y. et al. (2021). Computing power network: the architecture of convergence of computing and networking towards 6G requirement. *China Communications* 18 (2): 175–185.

10 Song, L., Hu, X., Zhang, G. et al. (2022). Networking systems of AI: on the convergence of computing and communications. *IEEE Internet of Things Journal* 9 (20): 20352–20381.

11 Parygin, D., Nikitsky, N., Kamaev, V. et al. (2016). Multi-agent approach to distributed processing big sensor data based on fog computing model for the monitoring of the urban infrastructure systems. *2016 International Conference System Modeling & Advancement in Research Trends (SMART)*, 25–27 November 2016, Moradabad, India, pp. 305–310. IEEE.

12 Saad, W., Bennis, M., and Chen, M. (2019). A vision of 6G wireless systems: applications, trends, technologies, and open research problems. *IEEE Network* 34 (3): 134–142.

13 Chowdhury, M.Z., Shahjalal, M., Ahmed, S., and Jang, Y.M. (2020). 6G wireless communication systems: applications, requirements, technologies, challenges, and research directions. *IEEE Open Journal of the Communications Society* 1: 957–975.

14 Chen, S., Liang, Y.C., Sun, S. et al. (2020). Vision, requirements, and technology trend of 6G: how to tackle the challenges of system coverage, capacity, user data-rate and movement speed. *IEEE Wireless Communications* 27 (2): 218–228.

15 Lu, W., Si, P., Huang, G. et al. (2020). SWIPT cooperative spectrum sharing for 6G-enabled cognitive IoT network. *IEEE Internet of Things Journal* 8 (20): 15070–15080.

16 Finkbeiner, T., Hush, G., Larsen, T. et al. (2017). In-memory intelligence. *IEEE Micro* 37 (4): 30–38.

17 Siegl, P., Buchty, R., and Berekovic, M. (2016). Data-centric computing frontiers: a survey on processing-in-memory. *Proceedings of the Second International Symposium on Memory Systems*, 3–6 October 2016, Alexandria VA USA, pp. 295–308.

18 Ahn, J., Hong, S., Yoo, S., Mutlu, O., and Choi, K. (2015). A scalable processing-in-memory accelerator for parallel graph processing. *Proceedings of the 42nd Annual International Symposium on Computer Architecture*, 13–17 June 2015, Portland Oregon USA, pp. 105–117.

19 Park, J. Chiplets and heterogeneous packaging are changing system design and analysis. Cadence White Paper, accessed online on 1 September 2024. https://www.cadence.com/content/dam/cadence-www/global/en_US/documents/tools/ic-package-design-analysis/chiplets-and-heterogeneous-packaging-are-changing-system-design-and-analysis.pdf.

20 Tian, W., Li, B., Li, Z. et al. (2022). Using Chiplet encapsulation technology to achieve processing-in-memory functions. *Micromachines* 13 (10): 1790.

21 Li, T., Hou, J., Yan, J. et al. (2020). Chiplet heterogeneous integration technology – status and challenges. *Electronics* 9 (4): 670.

22 Orenes-Vera, M., Tureci, E., Wentzlaf, D., and Martonosi, M. (2023). Massive data-centric parallelism in the chiplet era. *arXiv preprint*. arXiv:2304.09389.

23 JEDEC Press Release. "JEDEC publishes HBM3 Update to High Bandwidth Memory (HBM) Standard." Arlington VA USA, 27 January 2022. https://www.jedec.org/news/pressreleases/jedec-publishes-hbm3-update-high-bandwidth-memory-hbm-standard (accessed 1 September 2024).

24 Shantharama, P., Thyagaturu, A.S., and Reisslein, M. (2020). Hardware-accelerated platforms and infrastructures for network functions: a survey of enabling technologies and research studies. *IEEE Access* 8: 132021–132085.

25 Macri, J. (2015). AMD's next generation GPU and high bandwidth memory architecture: FURY. *2015 IEEE Hot Chips 27 Symposium (HCS)*, 22–25 August 2015, Cupertino CA USA, pp. 1–26. IEEE.

26 Choi, Y.K., Chi, Y., Qiao, W., Samardzic, N., and Cong, J. (2021). Hbm connect: High-performance HLS interconnect for FPGA HBM. *The 2021 ACM/SIGDA*

International Symposium on Field-Programmable Gate Arrays, 28 February–2 March 2021, Virtual Conference, pp. 116–126.

27 Synopsys. What is high bandwidth memory 3 (HBM3)? Synopsys HBM3 Online Product Page. https://www.synopsys.com/glossary/what-is-high-bandwitdth-memory-3.html (accessed 1 September 2024).

28 Li, W., Chen, J., Cao, J. et al. (2022). EID-GAN: generative adversarial nets for extremely imbalanced data augmentation. *IEEE Transactions on Industrial Informatics* 19 (3): 3208–3218.

29 Yoon, J., Jordon, J., and Schaar, M. (2018). Gain: Missing data imputation using generative adversarial nets. *International Conference on Machine Learning*, July 10–15 2018, Stockholm Sweden, pp. 5689–5698. PMLR.

30 De Melo, C.M., Torralba, A., Guibas, L. et al. (2022). Next-generation deep learning based on simulators and synthetic data. *Trends in Cognitive Sciences* 26 (2): 174–187.

31 Lu, Y., Wang, H., and Wei, W., (2023). Machine learning for synthetic data generation: a review. *arXiv preprint arXiv:2302.04062.*

32 Cao, Y., Li, S., Liu, Y. et al. (2023). A comprehensive survey of AI-generated content (AIGC): a history of generative AI from GAN to ChatGPT. *arXiv preprint arXiv:2303.04226.*

33 Park, K., Mott, B.W., Min, W. et al. (2019). Generating educational game levels with multistep deep convolutional generative adversarial networks. *2019 IEEE Conference on Games (CoG)*, 20–23 August 2019, London UK, pp. 1–8. IEEE.

34 Morsali, A., Haghighat, A., and Champagne, B. (2022). Deep learning-based hybrid analog-digital signal processing in mmWave massive-MIMO systems. *IEEE Access* 10: 72348–72362.

35 Zhu, G., Liu, D., Du, Y. et al. (2020). Toward an intelligent edge: wireless communication meets machine learning. *IEEE Communications Magazine* 58 (1): 19–25.

36 Xu, W., Yang, Z., Ng, D.W.K. et al. (2023). Edge learning for B5G networks with distributed signal processing: semantic communication, edge computing, and wireless sensing. *IEEE Journal of Selected Topics in Signal Processing* 17 (1): 9–39.

37 Jagannath, A., Jagannath, J., and Melodia, T. (2021). Redefining wireless communication for 6G: signal processing meets deep learning with deep unfolding. *IEEE Transactions on Artificial Intelligence* 2 (6): 528–536.

38 Nam, J.W. and Lee, Y.K. (2021). Machine-learning based analog and mixed-signal circuit design and optimization. *2021 International Conference on Information Networking (ICOIN)*, 13–16 January 2021, Jeju Island South Korea, pp. 874–876. IEEE.

Index

Edge Computing Acceleration: From 5G to 6G and Beyond, First Edition.
Patrick Hung, Hongwei Kan, and Greg Knopf.

d

f

g

h

i

n

o

p

q

r

s

V

W

X

Z

The ComSoc Guides to Communications Technologies

Nim K. Cheung, *Senior Editor*
Richard Lau, *Associate Editor*

The ComSoc Guide to Next Generation Optical Transport: SDH/SONET/OTN
Huub van Helvoort

The ComSoc Guide to Managing Telecommunications Projects
Celia Desmond

WiMAX Technology and Network Evolution
Kamran Etemad and Ming-Yee Lai

An Introduction to Network Modeling and Simulation for the Practicing Engineer
Jack Burbank, William Kasch, and Jon Ward

The ComSoc Guide to Passive Optical Networks: Enhancing the Last Mile Access
Stephen Weinstein, Yuanqiu Luo, and Ting Wang

Digital Terrestrial Television Broadcasting: Technology and System
Jian Song, Zhixing Yang, and Jun Wang

TV White Space: The First Step Towards Better Utilization of Frequency Spectrum
Ser Wah Oh, Yugang Ma, Edward Peh, and Ming-Hung Tao

Digital Services in the 21st Century: A Strategic and Business Perspective
Antonio Sanchez and Belen Carro

Toward 6G: A New Era of Convergence
Amin Ebrahimzadeh and Martin Maier

VCSEL Industry: Communication and Sensing
Babu Dayal Padullaparthi, Jim A. Tatum, and Kenichi Iga

Intelligent Reconfigurable Surfaces (IRS) for Prospective 6G Wireless Networks
Muhammad Ali Imran, Lina Mohjazi, Lina Bariah, Sami Muhaidat, Tie Jun Cui, and Qammer H. Abbasi

6G and Onward to Next G: The Road to the Multiverse
Martin Maier

Evolution of Wireless Communication Ecosystems
Suat Seçgin

Edge Computing Acceleration: From 5G to 6G and Beyond
Patrick Hung, Hongwei Kan, and Greg Knopf

Printed and bound by CPI Group (UK) Ltd, Croydon, CR0 4YY
17/09/2026
14964252-0002